My Side of the Desk

A teacher's take on school, students and parents

by

Lynne C. Bolinger

teacher and columnist for *The Kokomo Tribune*

First Edition, 2001

Photos courtesy of :
Kokomo High School *Sargasso*
Kokomo Public School Media Center
Kokomo, Indiana

Manufactured in the United States of American

ISBN: 0-9715966-0-3

Dedication

Praise God from whom all blessings flow

and

Mike, who abides my whims
Allyson and Zack, who have taught me how to parent
Colleagues and students, who have taught me how to teach

Preface

Today's schools are different from what they used to be. Where a high school diploma was once the reward for roughly 50% of the school-age population, that slip of paper is now considered a necessity as students move into the adult world of life and work. Traditional curriculum and methods which were directed to the college bound, must now be amended so that students who might have ended their schooling early can now finish high school and prepare for careers. This doesn't mean that schools can water down their requirements. In fact, as more states demand some sort of graduation exam, school standards are rising all over the nation.

Today's students need know-how in technology just to survive in whatever career they find themselves, so schools now feature banks of computers in media centers and labs, where students can hone these skills. In nearly three decades of teaching English, I have been a part of these changes.

Wheaton College gave me a liberal arts degree and a love for literature. I had passed the requisite course work to get a license but it took on-the-job training, time in the classroom with real students, to learn how to be a teacher.

I have been blessed with colleagues who have (as needed) shepherded, counciled, reprimanded or cheered me as I've pursued my career in education. I discovered what most veteran teachers learn: we love teaching and we love kids. What more can you ask of life than a job where you can enter smiling most days?

As an added life perk, the *Kokomo Tribune* in Kokomo, Indiana gave me the opportunity to write for them on a weekly basis. What many teachers think, I get to express in print regularly. I have always loved to read newspapers and what an unbelievable gift it is to be able to write in this medium. The staff at my local newspaper has encouraged me in this area of pursuit. John Wiles and Tom Carey are my special mentors.

Today's schools face many challenges as we try to expand educational opportunities to all students, not just the academically gifted. I have been a part of the changing face of American education and find innovations exciting. On the other hand, some things never change: parents who commit to their child's education, who stay involved and provide what help they can, will see their efforts rewarded through the success of the student in their home.

Acknowledgment

Penny Hall, media specialist

Rachel Polk and Jim Sondgeroth

Jeremy Fewell and Zack Bolinger

Editors and friends at
the New Castle Courier-Times, Jeffersonville Evening News,
Frankfort Times, New Albany Tribune, Columbus Republic and,
of course,

The Kokomo Tribune.

The old Howard College sold its Normal Building to the Kokomo Board of Education in 1872 for one of its early high school buildings.

Chapter One

Advice from Teacher to Parent

Until they have their own kids, teachers can lay claim to being parenting experts. They witness good parenting every day. As teachers get to know their students, they learn about how they value themselves, their work and their play. Most kids value their parents' opinions, even if they don't say so. Parents have that potent tool to raise children into fine, successful adults.

We recently drove down to the big bookstore because we read that Kareem Abdul-Jabbar would appear to sign copies of the book he has written. I took a basketball fan to meet this man, one of the greatest to ever play the game. He is also a gentleman and a scholar.

The bookstore management anticipated a crowd, so they duct-taped a path on the floor that wound through their stacks. And they printed out a list of rules for the fans to follow:

> Buy the book before you get in line.
> Anyone in the line without a book will be asked to leave.
> Please leave the autograph area as soon as your books are signed.
> Mr. Abdul-Jabbar will only autograph books, including his new book, *Black Profiles in Courage*.

We assumed this famous man takes time out of his busy life to sell his book. That's reasonable. So what were all those people doing in line with basketballs and trading cards?

"Oh, this happens all the time," said the clerk as she rang up our books. "People bring in all sorts of things to get autographs. But, they're going to go home disappointed. He won't sign anything but his book. That's in his contract."

Everybody got a copy of the rules, printed on fluorescent yellow paper. Rules bring order to chaos. Rules help get things done. But some folks don't get it. There are always some people who read the rules but assume "This doesn't apply to me."

We were near the front of the line. Behind us, a couple buzzed about their strategy to get an autograph on their basketball.

"See those television crews? He can't make a scene." They would just plop the ball on the table in front of him. They figured that one look at their eager sons, and how could he refuse?

I was glad I was in front of them.

Another couple had a different sure-fire plan. They'd use their cute blond daughter to get a signature on a forbidden card. How could he turn down such a sweet little thing?

"Just smile and say 'please,' and bat your eyelashes, and he'll sign our card."

So, let's see. What lessons were Mom and Dad teaching? Rules don't apply. Use coercion to get what you want. Use your looks ... that's a girl's value...to get what you want.

The big man entered, took his seat, and the crowd surged forward. I thanked him when he signed our books. "My pleasure," he replied. We moved on toward the door.

But we would have had to be outside to miss the next few minutes. He politely declined to sign the basketball. He smiled at the cute little blond girl and politely declined to sign the card. The grown-ups pitched a fit, grabbed their children, basketballs and cards and stormed loudly out of the store. A rumbled moved through the line and several people dropped out.

It was unpleasant for the visiting celebrity, but also for the children. Were they not precious enough to impress the famous guy?

In school, we try to teach students to respect each other. Rules are part of that discipline. Surely, parents have rules at home that they expect their children to follow. But we teach more by our actions than our words.

These children are learning that their parents don't observe the rules. It follows, why should the kids?

Some parents are my heroes.

I was sorting through my morning mail in the main office when a young man walked in. Our students dress casually, so he stuck out like the proverbial sore thumb. He wore a crisp white, long -sleeved shirt tucked neatly into black dress pants. The pants were belted. The cuffs brushed on shiny black shoes. I detected a fresh haircut.

He was in the company of a tiny woman who towered over him when she announced, "I'm John's mother. He has been having problems with his behavior in class so I will be accompanying him to his classes this morning. I have to leave at noon and my husband will replace me."

The bustle of the office came to an abrupt halt. Visitors are common and welcome, but this was a first for many of us.

Mom asked for her son's schedule and then led him out the door. He slumped, hands in pockets and head bowed. But he followed right behind. They had obviously discussed procedure and consequences before they arrived.

The office buzz resumed and most went on to their usual day. One student, however, was going to get a unique experience in his education. He would walk into each class with his parent, make introductions to the teacher and explain that his parent was with him to help him behave.

When there is an academic problem in school, we can make suggestions about tutors, after-school help, and outside resources. When the problem is a student's behavior, it's a stickier issue. Kids often behave differently when out of the parents' sight. Sometimes, they behave better, showing off all the good training they've received at home.

Sometimes, however, they use the away-from-home freedom to exercise unacceptable patterns of behavior.

School officials suggest that parents come to school and follow their child to his classes. Parents often counter with the assurance that the bad behavior won't happen when they're present. Of course, then their child gets to see how much better life is when you toss the bad manners.

Other parents wail that their presence will embarrass their child excessively.

John's parents had decided that his bad behavior at school warranted him being embarrassed. Shame is a healthy, potent response to public rudeness and other forms of socially unacceptable activities.

When their child exhibits bad behavior, parents have to confront their own failure demons. "What have we done wrong? Haven't we taught him better than this?" This may be the hidden reason they don't want to come to school. But they should exorcise and get on with it.

Parents can take much of the credit when their children behave well. Obviously they've been busy training and teaching. But those same good parents don't have to take the rap when their adolescent brings shame upon himself. There are just too many variables, influences and choices a child can follow once he's away from the nest.

On the other hand, for their child's sake, for society's sake, parents cannot afford to throw up their hands in despair and allow the bad behavior to flourish. There are effective ways to tighten the screws and continue to teach that unacceptable behavior will be challenged and corrected.

Tagging along at school is one of those ways. John knows his parents mean business and that they expect him to behave. No one can teach that lesson as well as Mom and Dad.

Kokomo built its first high school facility, completed in 1898 on the corner of Sycamore and Market Streets in downtown Kokomo. The school burned down, probably the result of a faulty furnace, in 1914.

Mother's Day

The other day, I drove to school and caught all the red lights, so I got to know the people in the car behind me.

Mom was driving young teen daughter to school in their utility vehicle. Mom's hair was pulled back quickly, with straggling pieces hanging down. She looked like she'd just been called into service. A missed bus?

The occupants sat on opposite sides, as far away from each other as possible. Mom's eyes were on the road; daughter stared, sullen, out her window. Her arms were crossed over her chest.

Mom looked over and said something. Daughter did not respond. Mom said, "Answer me." (I'm a mom. I can read those lips in that scenario.)

Mom repeated this. Daughter blurted out a monosyllable. Mom raised her eyebrows, sighed, and drove on.

They reached the front of the school. Daughter got out, slammed the door, and didn't look back. Mom watched her little girl storm off to the building.

There's a lot of door slamming in front of our school on any given morning. It's tough to be the parent of a young teen.

Even if you know all that theory about how they need to break from parents and find themselves, it hurts deep down when you can still remember the sweet-faced child who brought you tea when you were sick and who now thinks only of himself.

Young teens are so preoccupied with themselves, they forget their gym suits and expect someone to make that special trip. They forget the research paper and assume someone will bring it to school. They forget their lunch money and reckon someone will get it to them. That someone is often Mom.

We do drive them crazy at this crazy time of their lives. We hope that all we've invested in them will some day grow a decent adult. We pray we've done our best.

Just what do these kids need and want in a mom, anyway? I asked my students the other day "Why is your mom a good mom?" They were eager to answer. Here's what they had to say:

"If I have trouble with my homework, she'll help me. She always asks me if I have homework. I do it right away after school. She makes sure I have all my homework done. She corrects me on my mistakes. She does what all good moms should: she worries about me."

"Mom takes me to all my after school activities. She brings me things when I forget them. She goes and watches my brother's baseball games even when it's cold and rainy. She spends her free time with me."

"Mom doesn't let me listen to music with excessive cussing or watch "R" movies for my own good. I know this."

"When I get in trouble, she doesn't just discipline me; she sits down and talks to me about it. She listens to what I say. Every day, she sits down and listens to how my day went. She always says and does the right things. She acts like a good friend who is smarter than my regular friends."

"What I'd like to tell my mom is how much I appreciate her. I don't always act like it but I do. I yell at her and bite her head off then I try to tell her I'm sorry but I can't. But, I love her. I really do."

When Mother's Day rolls around, I wonder what the chauffeur will get this year.

Call the Party's Host

We were out of town when the seismic 16th birthday occurred, so I had told my teenager that if he wanted to have a party sometime, just give me a few days' notice. He took me at my word, invited a flock of friends and announced that they'd be by on Friday.

He assured me that he would do all the work, including cleanup, and he would prove to be a man of his word. My job was to chaperone. But, I warned him, a teacher's presence can dampen the fun.

Several years ago, a friend's band was performing at a local tavern and he asked me to drop by. I walked into the main room, the party in full swing. Large groups of people were congregated at the bar, laughing and toasting. Some folks were enjoying the entertainment while they nibbled on Buffalo wings. A few had risen and were dancing.

From way over at the bar, someone shouted, "Hello, Mrs. Bolinger." BOOM. Then silence except for the music. Many eyes turned to me as they lowered their glasses, like they'd been caught doing something they shouldn't. After a long pause, they returned to their friends, giggling and muttering self-consciously but a pall now hung over the room. I slipped out as my friend mouthed a thank you.

All the guests on our list know me as a teacher. Many know me as their teacher. I decided to make myself scarce in my study.

I warned the host that I'd make a scene only if something bad happened. What I know about modern teen parties, I've learned from my students. Sometimes, somebody uninvited crashes; sometimes, kids find the liquor cabinet or bring their own. Occasionally, somebody smuggles in drugs. Not unheard of . . . couples get lost and end up in a bedroom.

None of this happened at our party, but it could have. Of course, it's less likely with chaperones around. I was surprised, however, that of the fifty kids at our home, only two parents called to see that adults would be present. Only two.

Now, several guests live nearby and their parents could see that we were home. But most kids attended without any adult checking on the chaperone arrangements.

Of the two who called, one is a friend who knows us well. He wanted to double check. The other lady wanted to apologize in case we were offended. No, I told her. Good for you. Never apologize.

Many of the parents who didn't call would say that they trust their kids. We all want to do that. But, since good parties can turn bad, wouldn't it be comforting to know that an adult is underfoot?

Several years ago, two disturbed young men in Colorado were making bombs in a garage. They were stockpiling weapons and plotting destruction at their high school. As the dust cleared in Littleton, parents across the country shook their heads, sure that they'd know if their kids were up to something.

But, here were forty teenagers whose parents may have known where they were. Or maybe not.

Two young guests made a point to find the hostess and greet her. Ah, parental training. Of course, you can't expect kids to remember all that stuff when they're out of your sight.

So call ahead when your kid goes to a party. Always call.

Permission

I've lived long enough to have experienced some history, so I'm a codger who I can entertain my students, between punctuation practice and capitalization fun, with tales of long ago.

The other day I was relating how my older sister used to complain because I got away with so much. In the mid-sixties, she wanted to get her ears pierced, an uncommon but fashionable trend among her peers. She begged and pleaded, but she had to wait until she was a senior in high school for my parents to give her permission.

She was justifiably ticked off when, two months later, the same parents said I could get mine pierced; in fact, my mom performed the operation. I had just finished my freshman year.

Older siblings say it's often that way. Parents set up strict rules for their older children and ease up as they get old and tired.

In my case, my mom likes to say that when I used to ask for permission, I wasn't really asking whether I'd do something. I was asking if I'd do it with their permission.

What an ornery kid. By the way, when I chose to do something without permission, I always got caught and punished. Such was my job, to teach my parents how to parent.

My students think it's funny that my sister and I made such a big deal out of ear piercing, but there's the history lesson. Those little holes used to make an edgy statement, on the verge of rebellion.

Many of my students are also tickled that in the olden days, kids had to ask for permission from their parents. Their parents might argue the point, but it seems more and more that parents don't require that their kids obtain permission.

Many parents today let kids make all their own decisions. School people hear it all the time. Johnny can't get to school on time? "He's responsible to get there." Mary doesn't do her homework? "That's her choice." Frederick disrupts, smarts off, won't behave?

"When he's at school, he's your problem."

It certainly makes that parenting job a lot easier, when you never have to say "No" or put down your proverbial foot. But, kids aren't capable of making all their own decisions; that's why God invented parents. And if parents want their children to succeed at school, many need to become more active in setting limits, in establishing guidelines, and in supplying consequences.

One of my students is currently arguing with her mom about getting her tongue pieced.

Ouch.

Ouch!

Her mom says no. Girl pouts.

Now, I dislike pouting teenagers as much as anybody, but that's an expected response when they hit the wall of parental limits. Nobody ever died from pouting.

Kids do better when their parents are actively involved. Parents who are out of practice could begin by taking back school decisions from their children. They'll be amazed that the world will keep spinning and the kids will be okay.

Stutter

If you've ever listened to a group of young children straining in conversation, you may have noticed that it's common for them to tumble over their words. Their brains move so much faster than their mouths, they trip and stammer as they try to be heard.

You might give it little thought unless one of those talkers is your child. Then, you might slip into worried mode. "Is my child developing a stutter?" You're wasting your fret with young children. Speech pathologists don't step in before the child is eight or nine years old. Before that age, talkers are still developing their speaking skills.

Stuttering is an interesting disorder. Many have studied it but few have any firm theories about its causes. Some of the professionals who work with stutterers are, themselves, stutterers who have learned to control this problem.

Stuttering is more frequent in societies that demand rigid codes of behavior or where society demands that speaking skills develop early. People who stutter are neither more or less intelligent than the general population. The disorder occurs across a wide range of personalities. Four out of five stutterers are boys. There is a strong genetic link but it's rare in oldest children.

It's not true that anxiety causes stuttering, but stuttering can cause anxiety. So if your child stumbles over his words, there may be things you can do to guide him past the stammering stage when he's young and help him develop into a fluent speaker.

When my son was about three, he was trying to crack into the rapid-fire conversations that flew around his house. He would often stumble over his words and he took a long time to get them out.

One day, we had occasion to stop by a kindergarten room. During the course of the meeting, the teacher addressed a question to my son. As he began to form an answer, halting, thinking, speaking, I observed how she reacted. She bent over, looked him directly in the eyes, and waited patiently for him to finish. It was as though there wasn't anyone else in the world except this eager speaker. She waited until he was finished and then she answered back to his comment.

She'd say she was just being polite, but she was modeling how we can help young children who stumble over their words.

If your child has some difficulty with his speaking, instead of wringing your hands, try this. Talk to him with a relaxed rate. It helps to speak slowly. Stutterers tend to talk quickly, to get words out before they block. Show him that slow speaking works well.

Talk with a normal decibel level . . . not loud, not soft. Use your usual conversational voice.

When your child speaks to you, maintain eye contact. Don't look away like you're frustrated. Don't stare with a fixed gaze, like you can't wait for him to finish.

As your child talks, listen to what he says. Then, answer, responding to his comments, not his stammering. Let him enjoy the back and forth of conversation.

In most cases, when children start school, speech problems clear up all by themselves. If your child's stammering continues, you can have him evaluated by the speech pathologist who will direct you to the help you need. Many times, it's right in his school.

Too Many Toys

The baby, surrounded by all sorts of educational toys, plays with an old shoe. The older child trips over his roomful of the latest movie tie-in gear as he wails that he's bored and has nothing to do. It's a cliché and a joke, although less funny to those who purchased all those do-dads.

When we buy fancy playthings for our children, we're doing more than spoiling them, according to Dr. John Rosemond in his book *The Six Point Plan for Raising Happy, Healthy Children.* We're inhibiting many of the valuable skills they need for success in school and in life.

In earlier times, children were left to make their own fun. Today, instead of expecting children to entertain themselves, we load them with the toys they want and the ones we think they need. Instead of sending them outside to play, to make up their own games and create their own adventures, we give them the "tasteless, odor-free counterfeit of the real world on television," according to Rosemond.

The work of childhood is play. Through play, a child can develop many of the skills he needs in his adult life.

What skills does a child need? He needs his imagination, his creativity, initiative, and resourcefulness. We want him to develop these parts of his brain. Then, we turn around and, perhaps unwittingly, we inhibit his abilities.

Take, for example, the latest action figures for last summer's blockbuster movie. Let's suppose you buy them all. What can your child do with them? He can recreate the action of the movie. He may take the adventure a few steps beyond, but for the most part, he can play only one game. Then he's done. He's bored. And you're steaming before the credit card bill comes due.

On the other hand, if the child still has his imagination working, you could give him an empty appliance crate. He might make a house, a fort, a space shuttle, or the treads of a tank crushing the enemy in the backyard.

If I had the neighbors over and my baby was gnawing on a paper towel roll instead of a pretty teething ring, I might be embarrassed. But, if I let the baby pick his own toy, he might go for the cardboard.

Dr. Rosemond says that some parents buy those expensive, bright-colored plastic, mass produced toys as a status symbol. But, he also says they've been tricked into thinking they are aiding their child's intellectual development. After all, many of the most expensive toys are labeled "educational."

He believes kids were better off when Mom sent them outside to make their own fun and supplied limited toys, reserved for rainy days. He says that the best toys for children encourage fantasy and have the ability to change, to become many things.

As a practicing psychologist, parents often seek his help with their child's behavior problems. If he thinks boredom is at the root, he'll suggest his favorite toys as a solution: oatmeal boxes, popscicle-sticks, a shoebox, buttons, pots and pans; also clay and finger-paint are high on his list.

Outside, there are rocks, pine cones and "glorious mud."

His criteria for a good toy: it must be capable of being many things; it should encourage manipulation; it must be age-appropriate and durable.

Does the world need a doll which simulate every activity of a real baby?

Maybe it's time to clean out the toy chest.

Meeting with the Teacher

In an ideal world, parents send their kids to school and everything runs smoothly. Teachers teach. Students learn. Children advance through their books and lessons without any stress, growing wiser by the day. Periodically, a notice comes home, announcing the child's success, and everybody is happy.

The reality of school life is sometimes far from the ideal. Whenever a group of people has a stake in the same thing, in this case, the education of a child, there are going to be disagreements on the best paths to tread.

As is the case with many disagreements, when parents and teachers don't agree, it's often because they've not communicated clearly. It's best for the student if these important people can unite in their efforts. For teacher and parents to see eye to eye, it works best if they can meet face to face.

Parent/teacher conferences afford a good opportunity for building a bridge and working together. Parents can call the school and schedule a conference with their child's teacher. If a school is committed to student success, the teacher will make an effort to meet with parents whenever she can, and as soon as possible after the parents make the request.

Few of us have time to waste. By following a few simple steps, you can make the most of a meeting and solve most problems.

First, if you have questions or concerns, schedule a conference as soon as you can. Don't put this off, as problems tend to grow. Request that at least one other member of the school's staff attend the meeting. This will keep everybody on topic.

Prepare by writing down what you want to discuss. Then, look it over and decide what is the most important. Arrange your concerns from the most pressing to the least. That way, if the primary topic takes much of the time, you'll still get it on the table. Your notes afford another advantage. If you go in with a script, you'll be less likely to stray and waste time.

On the day of the conference, call the school. Make sure the principle players are present. If the teacher came down with the flu the night before, she may be home with a splitting headache and your appointment could slip her mind. Teachers don't have secretaries who can call and cancel appointments.

Assuming all the parties are ready, get to the meeting on time. You may need a few extra minutes to park and find the conference room, but if the teacher has scheduled 20 minutes, you want every one of them.

After everybody meets, take out your script and begin. State you concerns. Tackle these one at a time. Then, listen to the teacher's response. As she talks, if you have questions, jot them down. Then, raise them. Again listen. Everybody at the table wants the same thing, for your child to succeed in school.

You can resolve most school problems with a simple meeting. When teachers and parents see each other as allies, the student they share will be the winner.

Quality communication is one of the toughest elements in most relationships. If you step back and analyze any problem between two or more people, mixed up messages are usually at the root.

Teenagers and parents often find that it's nearly impossible to chat, let alone delve into ideas, feeling, and values. That growing up thing keeps getting in the way. Kids are busy breaking away and parents are trying to let them.

It's as common to find that parents don't get to hear what their children are trying to say as those children find they can't talk plainly to their parents.

So, that often places teachers in a special privileged position, to field a child's comments that he'd like to direct to his parents.

Many education studies indicate that parents need to get involved in their children's schools, as one way to help their kids succeed. Some parents make this a priority when the kids are small. Elementary schools can usually find cookie moms and field trip dads to help out at school.

But, as the kids get older, parents begin to feel that their help is not needed or wanted. Schools find it tougher to get parents to help in middle school and almost impossible to snag volunteers in high school.

Part of the reason is that kids give their parents the idea that the adults are an embarrassment. I know whereof I speak. I've got my very own teenager. There are days when if I say it, it's an embarrassment. On other days, if I don't say it, that's an embarrassment. If I wear those shoes, they look stupid. If I don't wear them, what I wear is wrong . . . where are the other shoes?

So, we parents can get the idea that we are unwelcome intruders into our children's lives. But that's not true. Teachers know because we hear it. Most kids crave their parents' interest. That's how they measure how much we care, as we make time for them in what they see as our busy lives.

In response to a recent column, where I suggested that parents stay involved in their children's schooling, I received a note from a bright high school student. She had placed it on my desk, With her permission I share her thoughts.

"As a child and now as a teen, my parents have never really gotten involved at my school or really with anything that has to do with school. I never asked for help on homework and they never offered."

The student continued, "I think it's really important to a child that parents do get involved. I wish my parents would have. Maybe we would have grown closer together."

There are kids who come from homes where nobody cares. Teachers know about this, also. These children have the cards stacked against them and must fight battles every day to succeed. But, there are lots of kids from intact families who long for more active parent interest

My source could be any number of high school students, kids who have learned to succeed without much help. Perhaps she'll be a stronger person for it, but she longed for some support at home.

How sad for her; how very sad for her parents. They may never know what they have missed.

Ben Is Getting Ready to Read

My family tends to sit in the same place every Sunday morning in church. Our pastor would prefer that we change locations and meet new people from time to time, but there's comfort in that familiar pew. Since sermon themes may make us uncomfortable, we gravitate to our secure spot.

The plus side of this transgression is that I've been watching Benjamin grow up. He's an engaging little boy with dark, wavy hair, pink cheeks and bright eyes. He sits dutifully between his mama and dad, trying to be quiet. They have a regular pew, also.

Ben's out of diapers now; he wears big boy clothes. And he tries to sit still during the opening of worship. It's a chore. He is a bundle of energy and stillness is not in his repertoire. He likes to peek over his shoulder at us and then he grimaces as Mom taps him back around to face the front.

Recently, Ben demonstrated a remarkable skill. His mom handed him a hymnal to give him something to do. He opened it. He realized that it was upside down. He flipped the book around and began to leaf through it.

Ben knows his way around books. Someone's been reading to him. He can't read yet, of course, but he knows which way the letters and numbers go. He'll be ready for school and for reading when the time comes.

When you read the statistics about literacy rates in our nation, you can't help but get depressed. And if learning to read was painless for you, you get angry that we spend so much money on education and kids can't read.

A great many children are ready to learn to read when they reach six years of age. That coincides with the first grade for most of them. If they've been raised in a print-rich atmosphere, if they've received the kind of mental enrichment that stimulates their language-receptor areas of the brain, they will probably move onto second grade as readers.

But, not all of them. I know of at least one very bright person who wasn't ready to read until he reached third grade. His parents worried, his teacher fretted, and he went about his business, learning some of the class content, but he could not read. Then, when he was about eight, everything jelled and he was off and running.

One goal teachers have is that all children will start school ready to learn. Much of that preparation begins before the child can sit up.

As a parent, you can do a lot to prepare your child for reading. From the beginning, when you talk to your baby, when you respond to his coos and grimaces, you teach the beginnings of language. These interactions are so natural, you may not realize that you're teaching, but you are.

If you want your child to be able to read, the single most important thing you can do is read to him. You can start as soon as he can be balanced on your lap. There are many bright-colored, simple books that are chew-proof and bathtub-durable that make good reading for a baby. No, he won't read back to you. But, he'll grow accustomed to the lilt of language and he'll begin to understand complex-sounding things as syntax, phonemes and grammar.

Someone who loves Ben has done him a big favor, preparing him for the adventure of reading.

Don't Buy Into a Lemon

It's like a bad joke. We purchased a dishwasher that turned out to be a lemon.

The machine performed its little duties for two years and then started to fall apart each time we used it. By then, of course, the warranty for what was breaking had expired. And the store from which we purchased it had closed.

We've dealt with rust, rotting shelves, major valve breakage, haunted wheels that refused to stay attached and gaskets that wouldn't seal.

Leaks have ruined a spot on the wood floor and some dishes have been crusted by whatever melted when the gasket came loose.

Over the months, I developed a relationship with an honest repairman who told me several times, "It's just badly designed, ma'am."

I finally had the old junker (six years isn't so old) carted away and replaced it with a better model. I know this because my repairman-friend says he never gets calls to fix this kind.

I'm an impotent consumer who's been suckered. That feeling of powerlessness makes me angry, if I dwell on it. All I can do is write the slam letter: "Dear Sir, If someone offered me one of your dishwashers, for free, I would run in the other direction."

After the warranty expires, it's all a consumer can do.

I wrote a different letter recently, when I was treated badly by a local mid-priced restaurant. The manager was very rude to me, tried to bully me, and told me I was wrong. (The customer is always right, sir.)

Folks at the national office called me, expressed their regrets and asked how they could make me a happy customer again. I suggested a gift certificate; they sent one large enough to take my family there. And, at my request, the manager did not come to our table and apologize.

Sometimes as consumers, we feel powerless; other times, we can find a way to rectify a bad situation.

I'll never buy another appliance of a certain brand and I'll warn my friends. I may go back to the restaurant, but I'll not stay if the manager is present. I'm older and wiser.

As parents, we are consumers of the educational business. But, we cannot afford to learn, after the fact, that our children did not get a quality education. There are no second chances. There is no gift certificate, inviting us back in for another try.

If you're an optimist, you'll send your child to school and assume all will be right. If you're more pessimistic, you'll cast a wary eye at every note, worksheet and grade card. Either way, you need to stay on top of what happens with your child at school.

You cannot assume everything is just fine. It may be, it usually is. But, if your child fails math, for example, he'll have to do a lot to be able to face math the next year.

You are not powerless. You can get involved in your child's education.

When your children begin a new school term, get to know your child's teachers as soon as you can. If you can go to school and meet them, that's great. If that's not possible, send a note to the teacher, introducing your child and outlining any concerns.

Some parents don't know what's reasonable. A daily phone report is not reasonable. A weekly report is. Expecting the teacher to return your call is. Expecting the building principal to meet with you is.

You are not powerless. You have a lifetime warranty on your child. Step forward. Make us accountable.

Gifts from Auntie

Stores stock the check-out lanes with potential impulse purchases. We were buying supplies for a family get-together and I fell for it. I grabbed enough PEZ dispensers for all of the cousins.

"You're trying to be the favorite aunt," my sister-in-law charged. I guess she was right. This was one of those not-often-enough gatherings. I rarely get to see my nieces and nephews.

And the candy was a hit. The kids tore into them, inserted those little sugar tablets and started to chomp. Within minutes, it was all gone. Within the hour, the cases were lost or broken. There had been that momentary rush when they squealed and smiled. But, then it was forgotten. I'd lose my favorite status with the arrival of the next group goody.

A similar thing happens each Christmas as we select gifts for the children on our list. We tend to want that squeal of delight. It's a high. So, we search for the current popular toy that all kids want. We'd like to wrap that and put it under the tree.

Then you watch the kiddies have at it. They open one gift, shriek, turn it over a few times and then toss it onto the pile of carnage. Next.

As we dive into the holiday season, stores will offer you any number of brightly colored plastic things, battery operated flashing things, and popular-movie merchandising things as gifts for those children on your list. I saw one today that screams everytime you squeeze it. You'd only buy one of those if you really dislike the child's parents.

If you'll take just a minute to walk to the other end of the mall, you'll find a store where you can buy gifts that will last. I mean, of course, the bookstore.

When I was eight, one of my aunts gave me a book for Christmas. I know I wrote that obligatory thank-you note, but I wasn't all that excited by the gift. There were lots of other toys to occupy my time.

After I'd played a while, I picked up the book. I opened it. I read it.

I can't recall anything else I got or who gave them to me. But I remember my aunt bought me that book

You might consider the gift of reading this year. Every time a child reads, his brain grows. That makes more room for learning. There are so many titles you can choose. Books cover topics from science, history, geography, biography, as well as fiction. Many compete with electronic media by using bright colors and 3-D illustrations. When children open your gift, you might not get that initial squeal (of course, you may), but you'll be giving something they can keep forever.

One tip before you go. Call the local school and talk to the librarian. Ask for his or her suggestions about literature for the child. Sometimes bookstores have lavish displays and will guide you to them. Sometimes, what they are selling is not what kids are reading.

Another good option is a subscription to a magazine. Most of these cost about the same as you'd spend on a toy that will be broken or lost all too soon. A magazine, on the other hand, keeps coming in the mail. Every month brings another present. The children's librarian at the public library can give you suggestions and the toll free ordering numbers.

If you've not written BOOK on your gift list, think about it before you go shopping.

The New Kid in School

Bring to mind, for a moment, the image of the lonely kid, standing at the door to his new school. Students stream in ahead of him as he shuffles slowly up the stairs. He wanders around, trying to find his locker and his class. All the kids stare at him and several whisper behind their hands. Then lunchtime comes and he sits all by himself. Later, he must get his books and get to the right bus before it pulls out of the parking lot.

Sometimes, kids have to change schools. A new job, a bigger house, divorce, re-marriage. No matter whether good news or a bad situation brings about the need to move, it's stressful for everybody. An added pressure is changing schools. And, starting into a new school can be tough on a child.

Kids tend to be more adaptable than adults but it can be difficult, even in the best of situations. A stressed-out student will have a tough time concentrating on his studies. Grade school children will blend in more easily than your junior high student, but there are some things you might be able to do to smooth everybody's transition wrinkles.

As soon as you know you're going to move to a new school area, try to prepare your children for the change. The more preliminary work you can do, the fewer problems the kids will have.

Help your children get acquainted with their new home. You can call or write to the Chamber of Commerce and request information about the city. Many community organizations will send photos of attractions, as well as maps and locations of historical sites.

When you contact them, try to connect with someone who has school-aged children. This person can tell you about what the local kids like to do during their leisure time.

If you can meet with school officials before the move, ask if your child's future teacher can arrange a pen pal for him. Perhaps several students or an entire class might take on the task of introducing your child to the school and the city. Then, when he arrives, there will be some students who feel that they already know him. They'll show him the ropes. The idea is to make him less of a stranger.

It also helps if kids can know about school schedules, bus stops, extracurricular activities and special programs ahead of time. They'll be less overwhelmed if they feel that they already know about their new school.

Teachers may tell you that it's best to move in at the start of a session. But, if timing a move by the school term isn't practical, don't fret about it. It's certainly easier for the teacher, but it's less critical for the child. At term change, he might get lost among every other new student. On the other hand, if he transfers in by himself, his newness will be obvious and the teacher might match him with another student to show him around.

What are your children's interests or hobbies? Try to locate organizations that he might join. Many churches have youth groups and children's ministries where your child can make fast friends. Because these groups are more social than academic, kids have lots of opportunities to talk and share. You might also check out 4-H and scouting organizations.

A new home, a new city, can be a fresh start for a student. Do what you can to smooth the way.

Chapter Two

Challenges Facing Today's Schools

As we enter the new millennium, society demands more from its schools. Just thirty years ago, high school graduation rates hovered around 30 % and that was acceptable for students and their parents. You didn't need a high school diploma to get a good job and carve out a piece of American pie.

But, now you do. Schools have redefined themselves as places where most students can succeed, not with watered-down curriculum but with newer teaching methods and different approaches to learning. We still face challenges, but that's a work in progress.

Circling the Wagons

Sometimes my little newspaper column strikes a big nerve and I get lots of feedback. That happened recently when I complained about my failing students who don't seem to care.

Teachers all over the place chimed in with their own tales of woe. I'm not the only teacher whose kids won't work. Lack of effort is wide-spread in middle and high school. Homework doesn't get done in affluent districts as well as in the inner-city. And everybody has a finger pointed at the person to blame

I was surprised to hear from many teachers that their principals blame them for high rates of student failure. Some administrators think the teachers should assure student success. Teachers get coerced into holding kids after school for extra tutoring. The principal wants his school to look good when compared to others in the district. All those F's are ugly.

Teachers blame parents, mostly. Once the student leaves the classroom, parents need to take over, say the teachers, who can't follow their students home. Teachers think parents should be more concerned with what the kiddies are doing after school. When a kid says, "Hey, after school is MY time, not yours," teachers assume he's picked up that attitude at home. Teachers wish parents would value their child's schooling and would make their kids do their homework.

Parents, many of whom want their kids to succeed, blame the schools and its administrators for setting standards so high and for insisting so much learning must take place outside of the classroom. They complain that their children come home with too much work and that often the parents cannot help their children because they don't know what the assignment is.

Missing in this blame game is the primary target—the student. It's time for the grown-ups to circle the wagons and attack the problem where it lies, in the center.

We all need to teach our students and children that they must reach for high standards and that there are consequences for not doing their work.

If a child has average intelligence and no learning disabilities, he should be able to earn a C in any required course if he's willing to work. Without help from parents, tutors, or even extra time with the teacher, he should be able to get that C. And with a little mental sweat, he could earn a B or maybe an occasional A.

I know, I know. Some kids have trouble outside of school, have a poor family life, live in less than desirable places. But, when these kids are out in the real world, will their boss care about any of that? No. The boss will want someone who knows how to work and who knows that if he doesn't, it'll cost him.

If a student doesn't do his homework, we need to place the blame squarely on his shoulders. If he says he can't do the work, we need to know that he can. If he says he won't, teachers, parents and administrators need to work together to convince him he will. Or else.

1951 Sargasso

High school students are deep in study for their next classess.

The Dumbest Thing I Learned in School

Now that you're an adult, out in the real world, here's a parlor game for you. Ask friends to recount the dumbest thing they were taught in school; it's probably something they haven't needed in their real-world lives.

For me, it was the "Cheery Breakfast" lesson in 8th grade home economics. Our teacher and our book stressed that this MOST IMPORTANT MEAL OF THE DAY must begin with everybody in a happy mood. So, as the cooks, we learned how to garnish hot cereal. We floated chocolate chips in farina, planted cinnamon sticks in rolled wheat and made smiley faces in the oatmeal with cling peaches. Oh yes, and we were to always wear a pink ribbon in our hair when we glided to the table. There were test questions over this.

In reality, the best breakfasts in my home are eaten in shifts. And I don't think we have a pink ribbon in the place. Don't get me wrong. There's plenty of necessary information you can learn about the economics of running a home. And those principles can carry over to running your daily life.

Unfortunately, I never got to take any of those classes because I was on the college-prep track and there weren't many open slots for electives. High school girls in the general track took Foods, Home and Family Living, Child Care, Home Finance and Marriage Preparation classes. There was a reasonable expectation that many of these ladies would graduate, marry and stay at home. In today's world, most 18-year-olds at home with babies are on the poverty track.

General education for boys sent them to vocational classrooms to learn a trade. The other day, while having my car serviced at a local shop, I discussed educational changes with the manager. He said that it's harder and harder to find workers who are competent to fix automobiles.

"When I was in high school," he said, "if a boy got in trouble, or wasn't very good in the classroom, he could at least learn to be a grease monkey...he could learn to fix cars." Then he swept his arm across the auto bays. "Do you see any grease here?"

In fact, what I saw were six or seven workers dipping in and out of the engines of cars. Each worker wore a white, short sleeved shirt. It was near the end of the day and there was no grease on anyone's shirt. Cars get fixed differently today.

The manager then told me that later that evening, they'd be having a study session to get ready for a certification test. "My technicians have to read and understand this manual," he said as he pounded on a phone-book sized tome. "Then they have to take a test and score 90% to pass. If they can't do that, I'm going to have to search for more workers because they can't stay here. Company policy."

He said he wished they came to him with better work skills, like punctuality, but he feels he can overlook sloppy attitude if they can just be competent. He said he has no choice.

"I wish you folks could bring a group of students over here for a field trip so they can see what it's really like, working on cars. It's not what they think. It's quite a shock for some people."

And perhaps it will be a shock for their parents who still think it's possible for their child to "just get that diploma" and then land a decent job. Look at the classified ads. There are a lot of jobs available. But employers need workers with a work ethic and some basic skills. Teachers tell students this all the time.

Parents need to join the chorus.

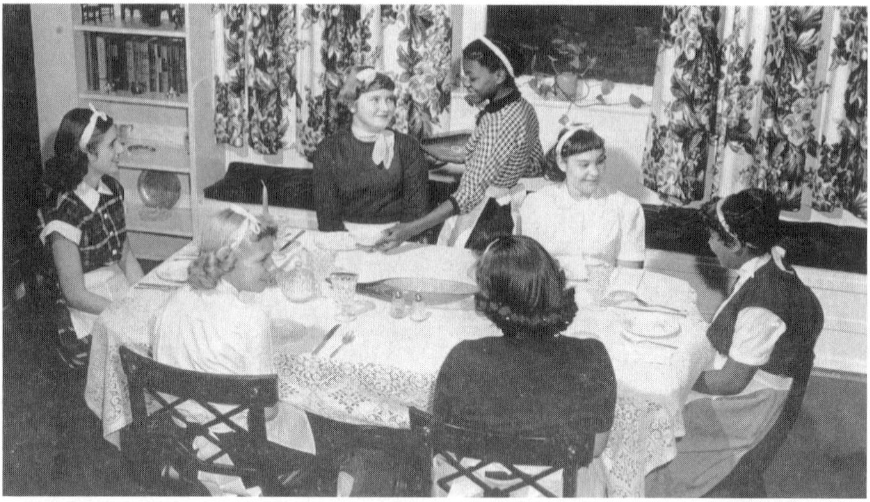

1953 Sargasso

Home economics students prepare and serve a meal with hair ribbons all around.

1953 Sargasso

Students learn the finer points of describing angles and arcs in the drafting class.

Freddie

Every teacher knows him. Let me introduce you to Freddie. He strolled into kindergarten one week late, holding onto his mother's hand. She explained that they had waited until kids were back in school to do their clothes shopping. "I hate the crowds," she said. "Anyway, it's only kindergarten. He didn't really miss anything, did he?"

In the next few years, Freddie's pattern was set. He was sick on the days when big projects were due. He came in late and the teacher had to refigure her grades.

Freddie also got sick whenever there was a test. He'd come back, expecting the class to wait for him. Teachers who tried to be fair gave him an alternative exam. Freddie cried that this was unfair to him.

According to our student handbook, no student can be absent the day before or after a scheduled vacation. So, Freddie was sick again. Phone calls to the office painted a story of a frail, little child. Or perhaps the caller claimed that Grandma or Auntie May had died. The child has a large family, with a great many grandparents. When Fred was a bit older, he bragged that his mom lied when she called in. As if we hadn't figured that out.

Now it's Fred. Fred fell and sprained his ankle. He hobbled into school, tardy of course, with a doctor's statement that excuses him from PE. He also may leave class five minutes early since it will take him longer to get around. The cast came off two weeks ago. He moves fast enough to visit with friends between classes. But, since the doctor's statement has not expired, he still demands to leave class early.

Fred's mom called the school, irate. "I just knew you people had no control over there." It seems that someone has stolen Fred's $200 jacket from his locker. Mom wants the police brought in. She wants a security check of our locker system. She wants to sue to recover the cost of the jacket. Why should she have to pay for our failure?

She could ask Fred if he gave his locker combination to anyone. It's against the rules of course, but Fred's getting pretty adept at thumbing his nose at rules.

She might also pop in and check the lost and found. Recently, our school donated over twenty heavy jackets to local charities. Left on bleachers, on floors and in classrooms, these jackets ended up in the big box in the nurse's station. They lay there throughout the winter months. Two weeks ago, every day, we read the announcement:

"Come check the lost and found. We have many jackets. After Friday, they will be given to needy families." No one came to claim a jacket. Not even Fred.

Mom called the principal to complain about "those teachers." Fred has fallen behind in his classes, what with all his illness right around spring break. None of his teachers even tried to help him catch up. No one would offer extra credit.

Poor Fred. I assume he's not taking science. Every science teacher in my building lets students retake tests. This is a great way for a student to learn the material and it ought to help him raise his grade. The only catch is that the student has to come in after school. Guess how many kids took advantage of that this last grading period? Not one.

Not Fred.

Fred doesn't turn in his homework. His teachers recently met with his mother. She's upset because her son may not pass to the next grade. Here at the end, she wonders what she can do.

"I just do not understand it," she says. "He tells me he has all his homework done and that he's turning it in."

I look at her. "He's lying to you."

She's stunned.

Literature Censors

A parent came to school, scrap of paper in hand, to complain about a literature selection. He didn't want his child reading such trash. Then, he handed over a scrawled list of objectionable works and their tallies...how many times each appeared in the book the class would be reading.

I stared down at the cold and clinical listing and marveled how some crusader had devoted so much time to counting. Modern literature sometimes contains language considered impolite and profane. The numbers were impressive.

Then I shook my head. Someone had distilled a favorite work of literature down to a word count. It would be like figuring all the 'the's' in a paragraph. You'd completely miss the point.

Most English teachers I know love literature and they love to light a fire under reluctant readers. They cherish their books like children. So, when someone objects to a favorite, teachers often get defensive.

This month's objectionable book was John Steinbeck's *Of Mice and Men*. It is one of the most challenged books in American schools. Although the author explores profound themes, his choice of vocabulary is simple enough to put the novel at 7th grade reading level so teachers often choose this book.

You may recall . . . your teacher probably selected it for you . . . the story of Lenny and George, two drifters drawn to each other through compliementary strengths and weaknesses. Lenny is a simple-minded man who possesses enormous physical strength; George is savvy, smart and small.

The title comes from Lenny's fondness for small furry things, like mice. He likes to catch them, keep them in his pocket and pet them. But he doesn't know his strength and he kills the mice. Then, like a child, he tries to hide what he's done.

George must be the brain for both men as they travel from work camp to farm camp in the dreary days of the Great Depression. Ah, but in the midst of their sparse existence, they have their dreams. Lenny holds on to his dream of his own rabbit farm, where he'll have lots of soft, furry animals to love.

The novel's title also alludes to a Robert Burns' poem . . . the best-laid plans of mice and men often go awry. The novel turns on an accidental murder. As their dreams fade, George must make a heart-wrenching decision. It moves many readers to tears.

Censors don't like the language these coarse men use. If they read the book, they might more strongly object to what appears to be a case for mercy killing. But, many censors don't read the books they object to.

At some point, you have to let your children be exposed to words and ideas that run contrary to what they've been taught at home. Readers bring to literature who they are. They filter ideas through their values. You have to trust that students can think through different situations without sustaining moral damage. Reading an expletive doesn't file it into the active vocabulary.

I strongly object to reducing a work of literature to lists and tallies of words. Check this out:

5 my(s)
7 me(s) in 146 words.
4 I(s)

What kind of egocentric tract is this? It's the 23rd Psalm, a prayer of comfort and peace, held dear by Christians and Jews, memorized, framed, and revered. The tally tells you nothing of its beauty and poetry.

When you zero in on individual words in any writing, you lose the meaning. Many of the books that are challenged in curriculum contain offensive words. Censors might look past that to the meaning of the whole.

Motivating Students

Teachers do all sorts of finish-up tasks at the end of the school year. We collect books, put away our personal things and, after sorting for next year, file all those lesson folders. And, of course, we record the final grades for our students.

A semester's worth of grades gives a record of how a student did in an individual class. I can reflect on how a particular student performed on certain tasks, by seeing what grade he earned each grading period. So can his parents.

I wonder about the girl whose grade dropped from a B to an F in six weeks. Although we worked on different things, such a radical change ought to alert parents to some problem. Just what exactly will she say when they look at her card, I wonder. Or will anybody look at her card?

Some kids will let their grades slide during the final grading period. And some parents will buy into their story that it's no big deal. But, it is, of course. It shows a carelessness that will extend to other areas in a student's life.

It's not uncommon, especially with less mature students, to find that someone raised his grade several notches. Perhaps he grew to understand that low performance only hurts him. Or, maybe an important adult placed a little pressure on him. Or maybe that adult decided to help him raise his grade. Whatever the cause, the student is now on his way to success in school.

Frustrating for me are the slackers who have figured out how to work the system. They flunk and flunk and flunk; then they calculate what minimum grade they need to pass and they attain it. I guess, for them, grades are a game. And they beat the odds, so hurray for now. How smug will they feel, however, when they present their transcripts at future job interviews?

I can spot the achievers, too. I had one student in class who worked so hard, turned in every assignment, but just couldn't quite hit the passing percentage.

I pulled him aside at the beginning of the final term and told him that I had seen how hard he had worked and that if he kept it up, I could give him his credit in this course because that's an option my school gives me.

He thanked me but said he wanted no favors. Then, he continued through the last grading period. He got a tutor; he did every available extra credit assignment. He never missed a day or an opportunity to raise his grade. Guess what? He earned a B. Maybe only he and I know what a struggle and an accomplishment that was.

I told him, privately, that any time in his life he needs a recommendation, he's got a golden one from this teacher.

When those final grade cards find their way home at the end of the term, let me suggest, parents, that you discuss them. School grades are only one measure of the child's work and ability, but they ARE a measure and need to be taken seriously.

If math grades are slipping, perhaps some summer tutoring is in order. If English grades have slid, a little required reading might help.

If you need some specific guidance, call your school's principal. He's around and he can direct you.

Seeing What We Want To See

It began during a family trip to the country. Perhaps the driver was near-sighted; perhaps he was lonely; or maybe, he just wanted to provoke the family. He roused us. "Quick. Quick. Look. Over there! It's a deer."

All passengers bolted up. Books and magazines dropped. Earphones popped out. And we looked where the finger pointed. At a distant herd of cows.

Lots of moaning and complaining. The driver shook his head and said, "It looked like deer to me."

We christened the sport as Deer Spotting. It has mutated into our inside family joke. On other trips to the country, we have single-handedly discovered many new breeds of deer: the rare and elusive spotted deer, the short-snouted, pot bellied deer, the egg laying deer.

When we play our game, we see deer because we want to. Guests in our car shake their heads, scowl, and roll their eyes.

Some critics of public school have their own version of deer spotting. They want to find conspiracies, ulterior motives, and menacing plots buried in what we are trying to do.

Textbooks are a common target. Even though, by law, parents comprise a large vote in selection, critics warn that schools are pushing some sinister agenda. What we're trying to do is prepare our students for the future. We'd hope that they can go to college, technical school, or into the workplace and find success.

A local school adopted a series of books for elementary grades, comprised of what kids call "chapter books" or real books, in the place of primers. The idea is to give children high interest books to encourage reading. They selected from a list of some of the best in children's literature.

The critics, to their credit, read all the books and then published their concerns. They nixed all books published by the Sierra Club as having too strong an environmental slant. They disliked books where characters came from broken families. They wanted to eliminate any book whose plot hinged on lying, cheating, or stealing, unless the violater paid for his crime.

They also dislike books that, in their opinion, over-emphasize historical mistreatment of minorities.

I showed the list of books to a friend who has read to her children for years. As she scanned it, she said,"I love that book," and "My daughter laughed and laughed," and "That's so special."

One example is *The War with Grandpa* by Robert Kimmel Smith. The critics don't like it because they see the plot hinge on a child's disrespect toward parents.

The main character is a young man many students could identify with. Grandpa must come to live with his family and he is forced to give up his room. Is he mad at his parents? Does he blame Grandpa? Would he like to strike out? Yes, yes, yes.

He plots ways to make Grandpa miserable. He's mean about it. He feels no remorse.

But, Grandpa is no slouch. He starts his own campaign to get even. He's successful. In the end, they make peace. And the main character realizes how much he has in common with his grandfather.

The critic slammed the book because it was felt that Grandpa should have reported the child's misbehavior to his parents. Then he would have been punished.

Oh, please! We all enjoy stories where authority's nose gets a tweek.

While you should try to stay on top of what your child is reading, don't let the alarmists make you too nervous. Give me a child with his nose buried in a good book any day.

1955 Sargasso

Seniors gather in the library to study and catch up on the latest news.

1953 Sargasso

The study of Latin got students excited about ancient Rome.

The Tale He Told

Most parents and teachers would agree that it's best for the student if all the adults in his life are on the same page. But, as in most relationships, parents and teachers don't always communicate effectively. Something can arise at school and if the parties don't talk it out, those problems can fester and grow into a situation that can no longer be resolved.

Here's a little apocrypha: One day, there was a small problem between Junior and his teacher. They exchanged words. His teacher, who knows he must win, won. The little problem was settled, although Junior came away with his ego bruised.

That night, licking his wounds, Junior called his buddy and began talking about the incident. When he spoke it out loud, it didn't sound so bad, so he embellished the tale a bit. Dad was walking by the phone and caught one side of his son's conversation. He quizzed Junior, who related to him a version of an exaggeration. Dad steamed. He didn't like what his son said the teacher said or did. He remembered what it was like when he was in school. These kinds of things didn't happen when HE was in school.

The next logical step for Dad would be to speak to the teacher. That could be handled the next day. But, Dad wanted to fume a bit, so he called his friend the principal, or his golfing buddy on the school board, or his neighbor who is another teacher in the same school, or an administrator who is a lodge brother. They patiently heard his complaints about the incident, which was now in its fourth version.

Whatever ear he managed to bend, he found an adult who was sympathetic to his poor son's plight. His audience offered their opinion and since Dad is their friend, they certainly saw his point.

When the complaint finally reached a high enough office to institute an investigation, here's what happened: Administrator called vice administrator, relating the incident and asking for action. Vice called principal or supervisor, relating that there was a problem and asking for action. Supervisor called vice supervisor, describing an angry parent and pressure from above.

About then, vice supervisor made it to the teacher's room for a chat. By now, of course, the initial incident was a week old. Within his classroom, the problem between Junior and his teacher had long been forgotten. The teacher had to think back to the problem and try to remember the details. As is usually the case, the teacher had a good explanation and shared it with the vice supervisor, who shared it with the supervisor who called the vice principal, who called the principal, who called the school board member who talked to the superintendent, who may call the parent back.

Parent wasted two weeks fuming about something of little consequence to anybody.

This is not an efficient way to handle a potential problem. Some schools make it a policy that if a parent calls with a complaint, he's told he must speak to the teacher first. After that initial call, he can find other ears higher up. It's the most orderly way to deal with disagreements.

Save yourself some grief. Honor the chain of command. Talk to the teacher first.

Team with Teacher

We do a lot of team-building and group activities at school these days. Where two heads are better than one, four heads can be even better and by working together, they learn other valuable skills, like solving problems with those whose ideas are different than one's own.

One skill is learning to gauge another's reactions. In a group setting, if you put forth an unworkable task, you can tell right away by the group's response.

When you give someone a task he cannot possibly accomplish, you don't get partial effort. You often get nothing.

Keep that in mind when you forge a most important team: you and your child's teacher.

It's tough to get that one going. We can't meet on a regular basis or without a lot of planning. We can't chat on the phone daily or bring new ideas whenever they surface. We can't read the other's reaction to our suggestions as immediately as if we sit, face to face. But, with communication and effort, we can work together for our mutual goal: helping your child succeed in school.

For example, some parents are so frustrated with their child's past performance, they ask for the unreasonable. "Call me every time my child fails to turn in his homework."

It's not going to happen. Why? The teacher has at least thirty students in her classroom all day. Some have 120. If it takes five minutes to call each parent after several hours of grading, you do the math.

Also, teachers don't always get the papers graded as soon as they hit the desk What if the teacher is absent? She may fall several days behind in grading. She may not catch up until the weekend. By then, possibly several assignments won't be done.

But don't despair. If we put our heads together, we can find a practical solution to your problem, one that will work for both of us.

Let's talk: has your child been irresponsible in the past with getting his homework done? Let's try this for two weeks:

We'll agreed that your child will hand his assignments to his teacher personally, rather than passing them in with everyone else. Or he'll bring the completed homework to the teacher before class begins so she can check to see that it is done. Let's put the bulk of the work where it belongs: on the student's shoulders.

If he doesn't keep up his end, I'll call you that day.

Then, after I tell you what I'm going to do, I want to know what you're going to do. Let's decide right here. I'd ask you to insist that your child turn in the assignment, completed, the next day. If this means he misses the ball game, the dance, the church youth group, so be it. If completed homework is important, and I believe it is, then we both need to be committed to it. I'm willing to put in the time if you'll put in the teeth.

Also, make it easy for me to reach you. Catching parents takes extra effort, usually after school or in the evenings. If you want me to call you, give me a secure number where we can connect.

Two weeks of this intense supervision should be enough to help your child establish the habit. Our group project will be successful.

Learning from the BMV

A few years ago, Indiana joined other states by requiring its high schoolers to pass a standardized exam to graduate. Students must demonstrate competency in mathematics and language arts. They got their first shot as sophomores.

Of the initial group, 46% failed to meet the standard in the fall of 1997. Since then, students have had opportunities for school-based tutoring. They got several more opportunities to pass the test.

You can hear the grumbling all over the place. The students and parents who view school responsibilities as time-clock-punching, moan that if the kid has served his sentence, he ought to get that diploma.

Advocates of special education students are mounting law suits around the exam requirement. Some have students who cannot read, who will never be able to read, and so will not be able to pass the test. They claim the requirement discriminates against these special ed. students.

And, of course, some teachers don't want to take on additional work, like learning how to reach some students in new ways.

Perhaps we could take a page from the Bureau of Motor Vehicles, another place where a cross-section of the community meets. Sooner or later, everybody shows up at the BMV.

During one uncharacteristically long wait....I held #65; they called #40...I had time to watch how they work.

First up was the initial line.

Everybody who entered took his place and moved toward the lady at the desk. She screened all comers.

She asked the woman in the white linen suit what she wanted. A title. She asked to see the papers. She smiled, gave the lady #68 and motioned her to the waiting chairs.

She asked the gentleman in the tank top that almost made it to his belt to hand over his title. She looked it over and said something to him about ID. He wanted to argue but she smiled, said she was sorry, and pointed to the door.

One gentleman got a bit agitated when she told him he had to get rid of his coke. She signaled to a security officer who came over and escorted the fellow out.

Nobody has a birth right to a license or car title. At the BMV, you walk their chalk or you get out.

By the way, they rotate that desk position every half day. Employees get to share in that job.

Once, when I lost my wallet, I discovered that to get a new driver's license, you must produce picture ID. Am I the only person who's picture ID is my driver's license? Even though I could produce a birth certificate, even though the clerk knew me personally, I still had to get a notarized letter to go with the photo in a school yearbook.

The disabled driver has a special marking on his license, as do those of us who must wear glasses to see. And don't we want drivers with special needs to take extra care when they share the roads with the rest of us?

So, why not set a standard for high school diplomas? The state probably needs to develop some sort of certificate for disabled students who cannot meet the mark. But, don't future employers have a right to know if a prospective worker can't read or add? As Indiana sets standards for graduation, this defines and increases the value of a diploma. That's good for all of us.

No Math Fear Here

"OK. First row. Up to the board." Words that strike terror in many mathematics students. Even with paper in hand and correct answers, the knees quake, the palms sweat and the brain goes on vacation. All in front of an audience.

Lots of students are scared to death of the subject, and fear is a low-down, no-good teacher.

In my high school there was a haven in the math hall. One teacher pronounced his room a safe place, where no one should be afraid to try.

The best teachers create an atmosphere and construct the learning experience for their students. My math teacher formed his group of freshman into a team who learned how to pull together, to count on each other, to share and to learn.

No one was afraid of failure. We charged into algebra bravely, feeling fit for the challenge because he convinced us we were. And he faced our seats, filled with giggling vessels of hormones; he grinned widely and invited us along for the adventure of our lives.

The teacher was all business, but my lasting memory is of his smile. He smiled with his whole face. His dark eyes disappeared in his laughter squint.

Each morning, we would sit, settle, and take out the hour's worth of homework from the night before. The teacher took his seat at the desk and looked to his right. We knew the drill. We began reciting our answers, one by one. When someone missed one—it had to happen—he lifted his chin, looked at the student, raised his eyebrows and ask something pertinent. "Did you change the negative to positive?" or "On step three, what do you have, $2x + y$?" Usually, the student would find his (and our) mistake, smile, nod, and we would go on.

I was amazed that my teacher could zero in on the error from behind his desk. I decided that he knew just about everything there was to know about algebra.

Sometimes, his question got no answer. Or sometimes, most of the class missed a problem. At that instant, he'd pick up a marker, click on the overhead projector, and work the problem for us. Twenty-six heads nodded. And I could feel my tiny brain synapses glow.

Just to keep us humble, he'd give this little lecture:

"Hey, you guys." (overhead click) "I know about this much math." He drew a dime-sized circle on the large field. "And you know about this much." A much smaller circle. "So we all have a way to go."

The teacher taught mutual respect with his example. No matter if the question was serious or silly, he responded as he would to a thoughtful adult. The one absolute truth: no one who tried was a failure.

He sensed some apprehension as finals drew near. "I'll tell you what, you guys," he said. "I'll let you have one sheet of paper for the exam and you can write whatever you want on it. Write some notes, some formulas. You can write really small. Maybe that'll help."

Like he had just come up with THAT idea.

And, of course, we dove into the project of creating legal cheat-sheets. We had to go back to Chapter 1 and work through the book. By the time we finished our papers, we had reviewed all important material. What a sneaky way to motivate us in the warm month of May.

It's important for teachers to make their classroom safe havens for students. They need to feel protected from ridicule so they can try, fail, try again.

I didn't learn this in any teacher's education class. I learned this sitting across from the best teacher I ever had.

Facing Down Foul Language

I was sitting in a fast food emporium at 5:30 AM enjoying a peaceful breakfast. The shop was uncrowded at that time of the day. It was quiet since those few who were up were either still sleepy or were reading their newspapers.

Then through the door came a couple of men, suited up for a day of hard labor. Their jeans were well-worn and they sported caps embroidered with their employer's name. I wouldn't have noticed them, probably, except they brought their meal to the table five feet away from me.

I shook out section A and turned away, having just gotten to the editorial page. But my neighbors' conversation was impossible to ignore.

They animated with their hands as they filled the air with some of the more staccato of profanities. That's a quaint term these days. We used to call them swear words, dirty words, words that got you a mouthful of soap or a smack on the back of the head.

I looked around at other diners. At one table, a group of well-dressed retirees were eyeing each other nervously and whispering in distress. Several others gathered up their trash and moved to the door. The waitress circulating with the coffee pot kept pouring, with a frozen smile.

But I was angry. I'd had my solitude invaded by this coarseness and I'm bold in the morning. I leaned over. "Excuse me," I said. "Would you keep your voices down? Some of us are trying to enjoy our breakfast." Their expressions exposed that they didn't have a clue as to what my problem was. They dropped the decibels and then mumbled back and forth, trying to figure it out.

You can't pick up a magazine today without reading about our increasing lack of civility. Manners? Ha. Courtesy? Huh? Polite conversation? Say what?

Television, that prevailing media, has pushed the envelope and has become a barometer for what we will tolerate. Trash talk shows have raised in-your-face public humiliation to expected norms for some people.

One criticism leveled at schools is that we fail to discipline our students. We fail to civilize them so they can join the adult world. Well, the same society that accepts verbal filth sends its children through the school house door.

Some students experience difficulty when they have to adjust their language for the polite society the school tries to enforce. But they can do it, for the most part. The worse thing any student has ever said to me was, "I hate you." She was screaming. It had been a bad day.

A quality newspaper won't print the harshest thing a parent has said to me. No big deal. All teachers have similar stories.

Within the school, we try to teach civility along with the lessons. When teachers model respect, students can copy it.

Many of today's students have been drenched in offensive language before they come to us. It's hard for them to understand why it's OK for Mom to say something, but not OK for the student to say the same thing. It's another lesson we try to teach. But, we're fighting an uphill battle.

One thing we can all do is call a spade a spade, as it were. We don't have to abide foul language in our children or from their friends. We can erect a dike and stand firm.

Chapter Three

Great Schools

Some schools are great. Others are less great. But all are working toward the excellence that our society requires.

Whether goaded on by state accountability or individual pride, schools across America are facing the challenge to educate all students to prepare them for life.

Within all schools today, there are pockets of greatness. We would all do well to examine what works and move it into our schools.

Detention Tale

Sometimes, they make us proud.

He walked into my room quietly at the end of a long week. We had had some little tiff and he needed to log thirty minutes of detention. We had already talked about it so there was nothing much more to say.

It was Friday afternoon. The sun shone through the window and Indian summer called to us. Grades were due, so I got to work.

Although we didn't chat, I could watch him across the room. I figured he couldn't have much to do, with this being the end of a term. He took out a textbook and opened it. Math. Of course. There's always math homework.

He worked quietly for a while. Then, he finished his assignment and closed his book with one of those "there, I'm done" sighs. He stretched his legs and looked at the clock. He had fifteen more minutes to go.

Experience with this age student prepared me for his fidgets, his pencil tapping, his dragging of the shoe back and forth. But, he did none of that.

After sizing up the time he still had to serve, he sat back up and flipped open his notebook. He turned to a section of vocabulary words and began to read them. He'd look at a word, mouth it silently, then look away and speak to himself about the definition or spelling.

He had decided that he wouldn't waste this time, that he could log a little study and could learn something.

Why is that worthy of note? Because it's remarkable. Here was an obviously self-motivated learner, one who was learning not only his words, but how to budget his time and how to make efficient use of a few spare minutes.

Someone has taught this young man that schoolwork is serious, that he needs to apply himself, that he needs to be about learning whenever he can squeeze it in.

Those lessons are taught at home, by parents who value learning, who model a good use of time, and who reward such efforts.

When he gets out of school, he'll step into the biggest contest of his life. Soon, he'll have to stand shoulder to shoulder with his peers for that spot in college, at the choice technical school, and in the job market. With him will be the slackers, those who complain whenever the teacher gives a weekend homework assignment. With him will be the dropouts and those who just squeaked by to graduate. No matter how many inches tall he stands, he'll raise head and shoulders above his competition.

I don't have this young man in an academic course, so I don't know what kind of grades he's getting this year, but I know that he's a hard worker and that he's focused on his learning.

I'd rather have a good worker than ten lazy geniuses any day. So will his future employers.

1953 Sargasso

Most students look forward to study hall to catch up with friends and to get homework done.

Drug Free Schools

At school, we want a drug-free environment for our students. There are lots of programs and educational plans to encourage kids to stay away from drugs. Schools spend on average $10 per pupil on prevention education. But, statistics show that what we're doing isn't working. That's because part of the team is still in the locker room. Parents must get involved in this fight against drugs.

A recent national study of the Drug Free Schools program, conducted by the Department of Education, makes it clear that too many children are turning to drugs.

I wonder how other people handle statistics like this: 18 percent of eighth graders and 24 percent of ninth graders report that they are heavy users of alcohol, meaning they drank more than ten times or have been drunk at least once in the past thirty days. Eighteen percent of eighth and ninth graders have used marijuana in the month before the survey and 5 percent reported being heavy users. Twenty-six percent of the older students smoke cigarettes.

I spend my days with eighth and ninth graders. They are so young and have so much potential. I'm supposed to prepare lessons and teach them, but they can't come close to their abilities if their heads are foggy or their brain cells are already fried.

Now, I know it gets tough to keep track of kids once they get their licenses, but it ought to be possible to keep tabs on students as young as this. Do their parents know what the kiddies are up to? Do they care?

In my community, our school takes an official, hard stand on drug abuse. All athletes, cheerleaders, and student leaders sign an agreement not to use illegal drugs. They can lose their status if they break this promise. Also, we have instituted a program where any student who wants certain privileges must volunteer for random drug testing during the year. There's a lot at stake for violators

One day, I was cruising around my local video store, trying not to run into anybody. I spotted a couple of student leaders who recognized me. They tried to steer clear, but we came close enough that I could smell the cigarettes and the alcohol on one girl. There is no way she could walk by a teacher or a parent undetected. So, who was waiting for her when she went home?

Most of us who think we are doing a diligent job of policing our children need to look at those numbers. 25% are abusing drugs. Who are these kids? They don't all come from bad homes where there's no grown-up around. Some are your neighbors.

If you want your kids to stay away from drugs, you need to stay on it. At some point, your kids will bristle when you call the hosts before sending them to a party. They'll hit you with, "You don't trust me." If you're not careful, they'll bully you into thinking there's something wrong with you for checking up on them.

However, there are a lot of adults who see very little wrong with underage drinking and smoking. Some have their own drug-dependency problems and may even encourage these as lesser evils.

The school wants a drug free environment for your child. But, you have him longer than we do. If you want your child to choose wisely, you have to set an example and you have to protect him as long as you can. That chaperoned party at a home where you don't know the parents . . . how can you be sure there'll be adults there? How can you be certain that the parents won't supply the booze or some weed?

It's not just "no big deal."

Music Ideas

Allow me to take you into a typical middle school for just a moment. It won't be too painful, I promise. Your major shock comes as you watch these squiggly, loud, clumsy children and say to yourself, " I was never like that." Then, on the heels of your pronouncement, you realize that yes, you were exactly like that. There's a high burn-out rate for teachers of these students; it takes a lot of energy to keep up with them. And the noise . . ..

Anyway, as we sit in the back of a classroom, the time nears for the passing period. There's a rustle of books and papers, backpacks zip open and stuffing begins. The rumble grows amid desks sliding on the floor. Noise reaches a crescendo and then, over the din BRRRRRRRRRRRR. An ear-splitting bell or buzzer sounds and the classroom belches out its contents into the hall. Now for four or five minutes, the student body becomes a great ocean, pulsing its way through the corridors. Lockers open and slam, books drop and get picked up. A girl sees her current best friend in the whole world fifty yards down the hall and greets her at the top of her lungs.

Then the tide recedes and students trickle off to different classrooms, doors slam, and again that loud bell sounds, a signal that class will now begin. Except, often we need several minutes to settle from the noise into a more quiet, learning mode.

Kids at this age are loud. They make noise when they move. Even when they're being good, they raise decibel levels. And teachers, sometimes forgetting that one adult cannot yell over thirty kids, try to talk over them. Noise, noise, noise.

The other day when I was riding in an elevator, I hatched an idea to cut down on the noise in this typical school. If I were running the world, I'd like to give it a try.

The lift was full and there were folks of various ages, including some teen-agers. As the elevator got to the floor, there sounded three soft dings. Ding, ding, and ding. People got quiet, waiting for the doors to open. Then they filed out.

I wonder what would happen if we replaced those loud school bells with three soft dings. Kids want to know when the passing period starts. Would they not quiet down so they could hear what would take an effort to hear?

Then to the passing period. Why not pipe classical music through the school's sound system while students walk on to the next class? I don't mean elevator music either, but even that would have a soothing effect as students naturally fall into step. But, why not some Brahms, Bach, Handel or Beethoven? I wouldn't have students select the music, because their tastes run to music that more closely mirrors their frantic, labile natures. But a school could spend a few bucks on some classical CDs and pipe it through the halls.

So students might begin their day to the trilling of trumpets. Later, they could waltz off to class with a Strauss waltz or glide to a Mozart aria. Then, at day's end, a Sousa march would lead them right out the door.

Our students might soak up a little culture without much sweat. And those adults who love these kids but hate the noise might last a bit longer.

Let's Hear It For Bus Drivers

We have a new motorist in our house. She recently graduated from driver's education and is awaiting the 16th birthday. For now, she must drive around with a parent as a passenger. Most of the time, I'm glad to give up my chauffeur's license.

On those occasions when she must slide into the other seat—recent ice and snow makes driving treacherous for anyone—she stays keenly aware of her new learning, most often by criticizing the less-than-fresh skills of her mother: "Ever heard of turn signals?" "I think you're supposed to park within walking distance of the curb." "Nice stop, Mom."

She's quite a fine driver, really, having learned from a team of experienced teachers. Her parents' job, in this interim, is to offer advice and pointers. As she drives us around town, she learns the routes to friends' homes and, of course, to the malls.

I've pointed out some dangerous intersections and road hazards. And I warn her to watch out for the Minivan Moms. You've seen them; perhaps you are one.

Overstuffed vehicles thunder past, and you can see loose children, dogs, and miscellany bouncing around behind the driver. You know the noise level in there is deafening.

My best advice to my daughter: "Never assume that a van driver is paying attention to what's in front or on a side street. Never assume she's seen you or that she has control of the pedals. Avoid whenever possible."

I'm graduating from the minivan ranks and I know whereof I speak. You can't concentrate on the road when someone behind you whispers, "Uh, oh. SHHHHHHHHH! Don't tell..."

So I gaze with wonder at those men and women who make their appointed rounds in the big yellow school buses every weekday morning and in the afternoon. The route is stop/start/stop, picking up one or two students at each station.

Minivan Mom may have a full house . . . seven passengers and God bless her. School transportation engineers gather forty to fifty as they complete their routes. Now, how much noise can that many children make, do you suppose?

Students' first contact with school is often through their bus driver. He can show by his face and voice that we're glad they've come.

But, social director is just a tiny part of the job. We entrust the lives of our children to those men and women. They must maneuver through dark streets, icy patches, and shifting snow. They must supervise rambunctious children while giving complete attention to road conditions and traffic.

They have to watch out for Minivan Moms plus all those people who are color blind to red lights and can't see stop signs. School bus drivers log over 1 million miles in a school year.

Just the other day, I ran an errand before school and got caught in a different route and traffic pattern. At one point, I was sandwiched between two yellow school buses, both pulling full loads. I watched as they moved carefully in rain slicked conditions. They did all the safe, regular things, like the full stop at the railroad crossing. Each motored with courtesy to other vehicles. Over the din. And, I know this is normal for our bus drivers.

They bring the children to school on time. And as your child climbs down, he's sent on his way with a smile and a friendly word.

Passion for Teaching

We were driving someplace the other day and my passenger was babbling. "Do you know how big Jupiter is?" No. He told me. "Do you know how many earths could fit in just the red spot on Jupiter?" How many? He told me. "Did you know that if I was standing on Mars and sent a radio signal, it would take a couple of minutes for you to hear it?" Yes, I knew that one.

He continued to dazzle me with his knowledge. "Guess how many Jupiters could fit on one side of our sun? Our sun is pretty small by galactic standards. You know how you can tell if you're looking at a star or a planet? Stars twinkle; planets don't."

Then it dawned on me. My son has caught his teacher's passion.

If you stood up in front of a bunch of teachers after a long school day and asked them about their passion, you might have to dodge a rice cake or a Diet Coke can. You'd see weariness as many lean chin into palm. You might hear groans.

Yet, these are passionate people. They prepare lessons and teach material, guided by curriculum and state competency requirements. They struggle each day with that ornery kid who wants to push. They endure the drudgery of paperwork, meetings, supervision, and the many other duties that go with the smooth running of a school. And it beats them down. But each has a passion, an area of knowledge that lights the fires. The school almost wouldn't have to pay the teacher when he gets to this part of the course.

A math teacher loves the intricacies of polynomials. There's nothing finer. Or quadratic equations. What precision! One teacher may soar when he shows students how to chart statistics and draw seven different graphs from the same data. Or he loves to demonstrate the practicality of size vs. volume.

One teacher's heart lies in the frigid North as the yearly Iditarod progresses. She loads her bulletin boards with silvery images of dogsleds. It's man against nature as her stories and pictures bring Alaska into the classroom. Her students hear the whistle of the north wind as they mush the frozen wilderness with her.

In one classroom, dinosaurs roam the earth again. Timelines and artifacts breathe prehistory to life. Across the way, the teacher preaches conservation and loves to take students outside to measure water quality. Other students pass geodes from desk to desk and explore nature's wondrous artwork.

Each literature teacher claims an author: Vonnegut, Poe, Hawthorne, C.S. Lewis, and Shakespeare. One wipes a tear whenever she teaches Capote's "A Christmas Memory." Some love to expound on a genre, like the epic, drama, or late 20th century American poetry.

Over in another hallway, students fight the Battle of Shiloh or listen for wild animals as they hike to the Pacific along the Oregon Trail. Perhaps they get to finger a World War medal or follow the track to Auschwitz.

From every corner of the schoolhouse, there's a little passion burning in each teacher. He warms the hands and minds of those students fortunate enough to be in his class. They learn also that knowledge is exciting and its pursuit is worthy of their best efforts.

The public school gathers faculty with diverse interests who can spark a student's imagination. If he's listening, he will catch the passion. If you're listening, you'll know.

Practical Teamwork

They keep dragging us into workshops to tell us about teamwork. Out in the real world, our students have to be able to work with people of different abilities; they have to learn these skills and society wants the teachers to teach them. Yeah, yeah, yeah, along with everything else.

Since a teacher doesn't have to venture outside his classroom, he may have trouble grasping the importance of the concept. But a quick trip into the work places of others demonstrates how teamwork is essential to success.

Our family recently spent a rainy Sunday in the emergency room of our favorite hospital. Suffice it to say that someone who should know better held a piece of wood in his hand and under a drill a second too long. Ouch.

He dashed in from the garage and shouted, "Hurry, we've got to go to the hospital." We slid into the car and sped down the street. The waiting room was filled with injured souls, all sizes and ages as it was home improvement season. Nobody dresses up for emergencies so we fit right in: I had on my crummy bike riding clothes and the injured one flashed shorts with assorted wood stains on them.

We had barely arrived when I looked over at my companion. He went from yellow to gray to white and said he was going to be sick. The desk nurse sat him down and took his blood pressure casually. When she charted 50\40, she glanced at me, smiled and said, "I think we'll just wheel him to the back so he can lie down."

After he was situated, she breezed back to me and said he was having a "vagal" episode and not to worry. "This happens all the time." She said they would hook him up to fluids, just as a precaution.

I signed the papers and joined my buddy in the back. They had him tubed and patched into several blinking pieces of equipment.

His vitals rose, they sat him up, he got dizzy, they laid him down.

"We'll do an EKG, just to be sure. We'll bring the X-ray tech here. We have coffee in the lounge; would you like some?"

RN Lori moved back and forth from the nurses' station to the cot, to the phone, to the doctor. Her voice remained steady and calm. It was near the end of her shift and they had been busy all day. She joked with the patient when he asked, "50 mg. of Demerol, please?"

"I'll put in your request," she said.

Kim, the certified respiratory therapist, filled in for the EKG operator who was gone that day. She explained what she was doing as she patted each sticky patch into place.

With all their explaining, I latched on to several words I'd never heard before. When I said I love new words, Lori brought me *Traber Cyclopedia of Medical Terms* to keep me busy.

RN Kathy got the worst job on this project. She explained, as she wrinkled her nose, that the doctor had to probe the wound to look for foreign matter. She had to stick several injections right into the hole. We all winced.

Next, the doctor smiled and stitched, finding no weird stuff inside the hand. He said that the patient was fortunate: all fingers wiggle and no bones are broken. He can still play his guitar.

Lori came back for the final clean-up and instructions. For the past three hours, I had watched a coordinated team, each with his own job, work together to reach the goal: a patient cared for and discharged from the emergency room. If one team member was absent or unable to do his task, someone else stepped into the gap.

These folks remain large and in charge in the middle of chaos. They see such a variety of injuries and illness; they have to be flexible, have to solve problems and have to keep learning.

Teamwork, problem solving, life-long learning: these are the skills we'll be trying to teach your children at school.

1917 *Sargasso*

Indiana high schools value their basketball teams. Here are members of the 1917 Kokomo High School squad.

School To Work

We were watching *Back to the Future* the other day. It's really fun when time-traveler Marty McFly lands in 1955 and gazes at relics of normal life. One of the biggest laughs comes when a Big Chrysler pulls into the Texaco station. As the ding-ding sounds, out scoot four uniformed men who descend on the automobile to service it. They buzz around frantically polishing chrome, wiping windows, checking tires and filling fuel tanks.

It's funny because you can't help marking how far we've come. Self-service is cheaper so most of us opt to fill up our own tanks. But it's rare to find workers who hustle, just because it's how a good job gets done. There used to be an ethic that drove employees to do their best because that's how they showed pride in themselves.

We've lost a lot of that. The other day, my students and I were discussing employment opportunities for underage workers. We listed chores, neighborhood jobs, paper routes, and errands they might do. We talked about how, since they don't qualify for most jobs, they could develop a clientele and hustle to earn extra money.

Then one perky boy piped in, "I wouldn't dirty my hands for a minimum wage job." We all looked at him. He'd completely missed the point.

"And how will you earn money before you are sixteen?" I asked. He smirked and said that between his mother, stepfather, stepmother and real father, there was plenty of cash for him.

He's in for a rude shock some day, but he illustrates an attitude we see more and more. Kids are growing up with no work ethic. We see it in school. We see it at their places of employment. When is the last time you were waited on by a young person who was obvious in his efforts to do a good job? Scratch your head awhile.

Pride in one's work—another value that parents used to instill at home. Some still do, but many of our students don't have any sense of this. Should we care? Well, yes, I think we should. Today's students will become tomorrow's workers. Our national quality of life depends on a quality work force.

Some students will go on to college after high school. Many will not or should not. More kids are heading for technical schools. These institutions are growing because they prepare students for the kinds of jobs that are out there.

Others of our students will leave formal schooling and go right into the work force. And employers are finding out, first hand, that students have no sense of what it takes to be a good employee.

School-to-Work, a federally funded program, is designed to give students the opportunity to explore the world of work. Local businesses agree to bring students in to the workplace so they can get hands-on experience. A paid employee allows the student to shadow him and learn what he does during his work day. It's a potential win-win situation. The student can find out for himself what a job requires. The business may be training a future quality employee for itself or for some other place of work.

It used to be that a kid with a diploma could walk into any number of jobs. This is no longer the case. The U.S. economy is becoming centered on the flow of information, according to the Federal Reserve Bank of Chicago. Their survey lists new skill requirements that apply to blue collar jobs, as well as entry jobs in offices.

Workers in all future fields need technical skills, the ability to communicate orally and in writing, and the ability to work in teams.

And many need to see what it's like in the world of work. School-to-Work is a program that will get it done.

1951 Sargasso

Many high school boys learned how to make furniture in wood shop.

1955 Sargasso

Secretarial students worked with brand new stenographic machines.

Teaching Teachers

A recent Fordham University report calls for better teacher preparation in Indiana. According to their statistics, this is an area where our state lags behind others.

I don't know. I have worked with three student teachers in the last few years, all of whom went on to graduate from Indiana colleges. Without exception, they were well-prepared, smart, and eager. My most recent colleague spent part of her sophomore and junior years working with difficult students in several inner city schools as part of her course work.

I'm not sure what constitutes good preparation for my profession. There are no better lessons than some sessions in an actual classroom with actual students. There's an art to preparing and presenting lessons, to keeping students focused, to following guidelines and meeting deadlines. It takes time and experience to develop those skills.

At school I'm surrounded by talented teachers who work hard to reach their students. Even so, grades and test scores indicate that some kids aren't learning. My neighbors are veterans who often still can't claim a high success rate. Whenever we send home midterm notices warning parents that a child may fail, the percentages are depressing.

However, if Indiana colleges want to improve the way they get teachers ready, there are some bits of knowledge somebody needs to impart to prospective educators.

How do you teach a student who comes to school without breakfast, without a warm sweater, without basic supplies?

How do you teach a student who doesn't choose to put forth some effort, who refuses to do his homework?

How do you teach a student who sees the teacher (and most other adults) as the enemy?

How do you teach a student who skips school or is chronically tardy?

How do you teach the kid whose big brother tells him, "It's OK. I never did well in math and I turned out just fine"?

How do you teach the kid whose mom takes him out of school on his birthday to go shopping because what's one day?

How do you teach the kid who decides that he needs to celebrate Columbus Day, Election Day, Valentine's Day and President's Day with a day off?

How do you teach the kid whose dad pulls him out early because "they never do anything " the week before Thanksgiving? Christmas? Spring break?

How do you teach a kid who hears at home that those school people better not take the beeper, cell phone, pocket knife, Pokemon card, cassette player away from him or else?

How do you teach a kid who sits and does nothing, learns nothing, then storms the desk during the last week, demanding some extra credit to save his grade?

Perhaps the Fordham folks can tutor future teachers so they can do a better job.

Hey. Sign me up, too.

Color-Coded Sense

My friend and I were rushing to an appointment and got stuck behind one of those sight-seeing types who slowed at every intersection and craned her neck to read the street signs. My companion finally pounded on her steering wheel. "What are you?' she screeched. "Somebody's grandmother?"

When we passed the slow car, we saw that yes, that's exactly what she was.

Then, my friend wheezed a guilty sign. She's not the kind to yell at somebody's Gramma. If only she'd known...wouldn't it be handy if we could identify the drivers around us on the road?

I'd propose a color-coded system. We could agree to purchase cars that identify us to each other quickly and accurately. We could divide up all licensed drivers by their demographics and driving records.

Senior citizens tend to drive slowly and their reflexes are delayed. That means that if you drive too closely behind them, they might suddenly hit the brakes and BOOM. Rear end collision. So, let's put them in white or gray cars and then steer around without anger.

Teen-age girls should pick black cars. When they park too close to each other at the mall, that color will hide the nicks.

Red must be assigned to teen-age boys. This will make it easy for the cops to spot them. When you'd see a red car approaching the four-way stop, you wouldn't even consider that he'll stop. You'd just sit there and let him fly through.

Oh, all right. Middle-aged men can have red cars, too, but they'll need to understand that cops will treat them like teen-age boys.

I'd make it a law that any kid-carrying minivan must be fluorescent and must have a strobe light that activates whenever the inside decibel level rises above 100. This will help the rest of us steer away at stop lights. The driver is already distracted.

For proven mellow drivers, those who don't measure their worth in terms of horse-power, blue is a genial color. You could approach them with confidence that they're driving like they were taught in driver's ed.

This leaves green and purple: assign those to trucks and sports utility vehicles and avoid trying to pass them since they won't let you.

This color system is a variation of something that works very well at school. On any given day, the teacher's mailbox is crammed with forms, memos, lists, reports, and other stuff that must be attended to. Some must be filled out and return, and some must be filed away. When we're lucky, they come in rainbow colors.

All those forms get stacked in the to-do basket and it's easy for some to get misplaced.

It helps when Mrs. Jones says, "I need that blue sheet back," or Mr. Smithe comes looking for the mint green paper.

And, although all forms must eventually be attended to, the teacher can avoid unpleasant ones until he's prepared. Those are usually fluorescent.

Color coding is a boon to those of us who need visuals to get things straight. What works at school could work on the streets around town.

A Clue About Your School

Listen to someone facing major surgery and when he talks about his doctor, he'll say, "and he's excellent." How does he know? He's talked to a few survivors, so the doctor must be competent, but excellent? What he means is, "Please, please, make him excellent."

It's the same thing for dentists, lawyers, and pastors — professionals in whom we place a great deal of trust not to hurt us and to do a good job.

And so it is with your child's school. When polled, many Americans will say that they fear problems in some schools, although a majority will say that those problems don't exist in their children's school.

How can you judge if your child's school is a good one? The obvious measure might be test scores and honor graduates. While these are important, there are many factors that are reflected in test scores, like the make-up of the student body.

Good facilities indicate that the community values the schools. That's a positive sign.

Another good measure is whether the teachers are happy. I know that sounds self serving, but the truth is there are a great many intangibles that happen within individual classrooms, and you want those peopled by happy teachers.

So, how does an outsider judge that? You might ask your child's teacher and you might get a good idea. However, if the teacher's having a bad year, you won't get an accurate reading.

Are you, the parent, made to feel welcomed when you enter the building? Does the staff help you by answering your questions?

When staff members pass each other, do they smile and say hello? What about how the staff responds to students they meet in the hall? You should see smiles and pleasantness. You should not feel a great deal of tension.

All these are hard to gauge. But here's one you can nail right away. If you have a problem with a teacher and you can get it solved without talking to the teacher, you have a school where teachers are not respected by whomever helped you.

If you call your buddy the principal or your neighbor the counselor, and he fixes the problem without bringing the teacher into it, he doesn't respect his teachers. Maybe you're dealing with a sticky problem, but if teachers are not respected, they are not happy.

What you'll find in excellent schools is a support system that extends to staff members as well as students.

A Case for Memorizing Poetry

Not too long ago, teachers routinely required their students to commit poems and historical passages to memory. Often the student would stand before his classmates and recite without notes.

Memory work has fallen on hard times. To some educators, rote memorization has been dismissed as less useful than more global learning. Many believe that it's more important for kids to learn how to find new information rather than cram their brains with lists of already known stuff. Prevailing wisdom is that since knowledge changes every day, it's less useful to commit to memory what may change.

Of course, there are some things worth memorizing like the order of the planets, the names of the Great Lakes, the colors in the spectrum and the order of mathematical operations.

And I'd like to make a case for memorizing classic poetry.

I reject the notion that all learning has to lead to a vocation. Sometimes, you can derive pleasure just from knowing things. Teachers see this all the time when the light bulb comes on in a child's face.

Not too long ago, an ancient student went one-on-one with a smart guy and took the prize.

My dad was engaged in a conversation with a man who had more formal education. The book learning guy was expounding on many topics, including the poet Robert Burns.

The more educated guy went on and on and on and then said, "...one of the best of the English poets."

A long pause.

"I believe," my dad began, "that you are incorrect. He was Scottish."

"No, he was English," said the book learning guy.

Another pause. This was a polite conversation over dinner and Dad wanted to gauge whether or not to make a point. He decided to go for it.

"Ah, no," he said. "Bobbie Burns, the great son of Scotland."

Then, from somewhere deep inside his memory cells, he brought forth a poem and recited it in Scottish. When he finished, a smile spread across his face.

Some long ago fourth grade teacher had required him to commit a poem to memory. His triumph gave him pleasure for a job well done so many years before.

William J. Bennett, former U.S. Secretary of Education and current critic of American education, says that rote memory work has its place in the learning of a student. He supports the memorization and recitation of fine literature.

Why should students be required to memorize and recite poetry? According to Bennett:

1. It's solid, demanding mental work and that's good for the brain.

2. Reciting before others makes excellent practice at public speaking.

3. Memorizing beautiful language helps children come to know it well. They can discover layers of meaning as they repeat them over and over.

4. Memorizing a few poems and passages actually attract many children to good literature.

Many adults still cherish the lines they learned in school. Today's teachers need to remember that and keep the tradition alive.

Chapter Four

Musings on Homework and Grades

For about eighteen years, the schoolhouse down the street looms large in a child's life. Its teachers become important influences in a student's character. And if only there was a magic potion that could help school years go smoothly. There isn't.

However, when parents take an active role in their child's education and when they work with the school, their kids will do well, will learn their lessons and prepare themselves for success in life.

Cure for the Common Scold

Your little angel is unique, I know. But most problems in school are common ones. Experienced teachers have solutions for most of the difficulties your child faces.

They will offer them to you with assurance that they will work. So you'll meet or talk over the phone. Most likely, you can work out a way to solve the problem. If, however, you're both perplexed or can't agree, here are some sure-fire solutions. They are simple and easy to monitor. You can find out quickly if the problem corrects itself.

Check through this list of everyday problems. See if anything sounds familiar.

- Student is not turning in homework. There are several reasons. Find out.

1. He does not understand how to do assignments.
Solution—He must ask the teacher for help, during study time or after school. She agrees to give him extra time until he can finish the work. If he is reluctant to ask for help, the

teacher needs to know that he's confused and needs to offer her help.
2. He's not doing homework at home.
3. He forgets to take it home/bring it back.
4. He tells you he doesn't have any homework or that he got it all done at school.
Solution—You and the teacher can help him get organized with little effort on your parts. Our school issues planners to all students. These have dated pages and sections for each class. If your child doesn't have a planner, you can buy an inexpensive spiral notebook that fits in his pocket.

You all agree that for the next three weeks, your child will write down daily assignments.

The teacher will initial the paper. So will you, when he's working at home. You have to supply consequences if he refuses to bring the paper home.

Then you'll check the paper and make sure he does the work. The teacher calls you on any day your child doesn't turn in his work. Give the teacher a "good" number, one where she can reach you without too much difficulty.

It's common for kids to have trouble with organization.

He needs to break the bad habit of forgetting his responsibilities. He may gripe, but he'll also be relieved to get you both off his back as he completes and turns in his homework.

- Student doesn't bring required materials to class.
  Solution—Get a list of what he needs to bring. Agree with the teacher on a plan of action; she can call you if he forgets again. Ask the teacher to help you—what suggestions does she have to help him?

- Student misbehaves.
  Solutions—communication is key. Together you and the teacher need answers to these questions:

1. Are class rules clear to both teacher and student?
2. Is the objectionable behavior acceptable at home? You might be amazed at what some homes consider normal. You might also be amazed at what some teachers find objectionable.
3. Who sits near the student? Changing seats is easy.
4. What strategies has the teacher tried and how have they worked? Here you can offer insight into the makeup of your child. Does he turn off his ears when someone raises her voice? A teacher needs to know this.
5. Is there something physically wrong with the child that the teacher needs to know? He might have slipped the medical screening. I once had a student who just could not stay awake during my edge-of-your-seat lesson on action verbs. We discovered she was experiencing petit mal seizures, which mimic daydreaming. She needed some medication to control the seizures.

Remember the old chestnut about catching more flies with honey? It's important to reward your child as he turns his behavior around, or as he becomes more responsible about school work. His grade will go up, and that's great. But many children need more tangible rewards. Only you know what that can be.

Most school problems can be solved when the parent and the teacher put their heads together. So, charge ahead and solve those problems.

How to Grow a Writer

Good parents want to give their children the advantage in school, and they've heard for a long time that one way to do that is to read to their children. Most feel competent to supplement what's going on in the classroom.

They fill the house with books and magazines, giving their child a print-rich environment. They restrict television and take the child to the library.

And most experts agree that reading to children is the single most important thing a parent can do to help his child do well in school.

However, another crucial aspect of helping a child in school is to teach about writing. It's the next step up. Learning to write goes with learning to read.

The only way for a child to learn how to write is to write. It's when he puts words down on a page that he begins to discover how it's done. Often, he'll learn this by trial and error. Each time you give your child a writing opportunity, you help him learn. Not all attempts have to be brilliant but each effort is a building block.

Writing is one measure of what the brain holds. It is becoming common for standardized achievement tests to include some sort of essay writing. Writers have to think about what they want to say; they have to find words to represent their ideas and they have to put the words on paper so it all makes sense.

Some writers make detailed outlines. Some brainstorm random ideas onto a page and then pick out the two or three they like the best. Although there are many routes, each writer has a unique method from ideas to finished product. The only way a writer can discover what works for him is to try, fail, try, succeed, try.

By writing, students gradually learn how to go about organizing and expressing their ideas in ways that can be understood by others Like many skills, writing requires practice.

But, how can you get your child to write?

Begin with the basic things.

Do you write out shopping lists? Have your child at your elbow, and write those lists together. Ask him what he wants from the grocery and then write his answers.

Ask him why he wants a particular item and jot that down, too.

Use a message board for family information. Try to write your messages as complete sentences. Don't worry if you're no longer sure how to do that. Remember that even faulty writing is a teaching tool.

How about thank-you notes? They're not only good etiquette; they are a gold mine opportunity to teach writing. Write these with your children until they feel some competence. Then, just give them the stationery and let them write them by themselves.

Ask your child to write captions for family photos. This will give him a chance to think and use his imagination.

Read what he writes and make comments. Many times, you can show him how well he communicates by telling him what you understand from his work.

Buy your child his own calendar and have him write his activities on it.

Don't be afraid of writing. Nobody's hovering overhead with a red pencil. Encourage your child to practice. The more he does, the better he'll get.

I Don't Do Homework

Everybody I talk to wants to improve our public schools. Study after study compares our students with those in other industrialized countries and we usually come up short. And in a typical American can-do manner, experts step forward with suggestions for improvement.

Most believe we should push our students, raise academic achievement and raise our standards.

On the heels of all this come pundits with policy suggestions and programs.

At their altitude, solutions need to be broad to accommodate our diverse population. But to those experts I'd like to issue an invitation to come into a real school and see what we are working with. Then, they can offer their solutions.

Let's enter a classroom in a typical school. Mine. We are closing out a semester and kids should have made the transition to the ways of schools. Yet, I look at a sea of failure—theirs and mine. At midterm, this group represented itself with fifteen Fs out of twenty-four students.

Like most veteran teachers, when results are so poor, I look for the cause: difficult material? unclear instruction? wide-spread illness? snow days? But, since I have a similar class that has a better distribution of grades, my vote: no, no, no, and no.

Next, to unravel the mystery, I turned to our newest technology and my computer grade book program. I can track how my students are doing in several kinds of activities. I checked their grades for in-class work and for homework. Guess what? They aren't doing their homework.

All of my F students have passing grades for the work we do in class. Many have 100% of those points. And many of them have less than 20% of the homework grades.

A quick assessment of assignments shows that some were longer than others were, but that didn't make a difference.

What's left? I asked them: what's the deal with homework? From the mouths of the babes:

(Giggle, giggle) "I forget."

"I thought it was due later."

"I couldn't remember which exercise you assigned."

Of course, if the student would write down assignments in a planner or a sheet of paper in his notebook, he'd know what he was supposed to do and when.

One non-scholar said, "Oh, I don't do homework," with the conviction of a life philosophy.

We chatted at length and I hoped I had motivated them. Then came the writing assignment that they had a week to complete.

Several of the Fs "didn't have time." In a week of seven days.

One young man, self-esteem firmly intact, turned in a brief paragraph that he whipped out during lunch. "At least I turned in something," he said. He was proud of himself.

All the programs, policies and platitudes won't raise student performance until someone makes students put forth quality effort.

Most teachers are concerned with student success, but here's the unvarnished truth: raising student achievement so they can meet higher standards will require work. Schools are working. Students need to join the team.

Learning Disablities and TV

In my first grade class, children were quickly placed into reading groups. We had cardinals, robins and canaries, but everybody figured out who the good readers were. For most of us, learning to read was so simple, it seemed automatic. And we were too busy to notice that the yellow group fell way behind the rest.

Some of these kids learned to read eventually, since maturity is one factor in reading ability. But many never did. They were doomed to school failure, year after year. They became the behavior problems, the slackers, or the dropouts. In earlier times, these children could still go on to find meaningful work in their adult years.

We didn't know about dyslexia, a disorder that scrambles signals to the brain. And the education experts didn't deal with children who learn differently. There was basically one way to get information and if students couldn't work that way, there was little for them to do but fail.

In today's public school, you'll find many programs for kids that address different ways to learn. At least 10% of the school population has identified language-based learning disabilities and qualify for special education funds. There are many more in regular classes who struggle with the printed word. Three out of ten students have real problems with reading and writing.

Today, we save more kids from failure. But, according to Dr. John Rosemond, psychologist and author, parents of pre-schoolers can do one important thing to help prepare their children to do better in school.

Turn off the television.

For decades, experts have been concerned about the content of television. Dr. Rosemond warns we need to be concerned about the process. He feels that the nature of the medium extinguishes the very skills we need to be good readers.

Learning disabled children share some of these problems:

They have visual screening problems . . . following a line a print across the page, and then moving down just one line; they lack attention-span, necessary for the skill of reading a line, then another line, then another line and getting the meaning; they lack listening and communication skills; they lack the ability to follow sequence: this, then this, then that; they lack hand-eye coordination and motor skills; they lack imagination and are easily bored.

Have you ever watched someone watching television? Sneak up on one of your kids and observe. You'll see the gaping mouth, the dull eyes and the motionless body. On average, the image on the tube changes every three to four seconds. The action changes back and forth, out of order with no logical sequence. And that flicker is addictive, sending him back for more. This develops a short attention span.

Even the adult television viewer gets his exercise by surfing channels with the remote. We seek that clicker fix.

The average American pre-schooler spends about 1/2 of his waking day watching TV. That's twenty-five to thirty hours weekly. Compare this to the child in Europe who watches an average five hours a week There is a much smaller incidence of learning disabilities in Europe and Dr. Rosemond says that's no coincidence.

He also says that parents make a mistake when they encourage their children to watch so-called educational television like *Sesame Street*. It's artificial to push letters and numbers on children and there's no research to indicate this early exposure is any kind of learning. Instead, children learn to expect a lot of flash, music and action.

What should they do instead of watching TV? Just about anything else.

Let's Take the Homework Problem

One Step Further

If you ask a bunch of teachers what stands in the way of student success, many will say the same thing: kids don't do their homework. Daily lessons revolve around last night's assignment, and the students who come unprepared aren't able to move ahead successfully.

So, I went to the source and asked my students. "Why don't you do your homework?" And for those who had no answer to that, "How do you manage to get all your homework done?"

Some responses show a lack of maturity. "Sometimes, I don't like my teacher, so I don't do his homework assignment." He wants to demonstrate his anger but he's not thinking. If each assignment takes ten minutes to grade, he's not hurting the teacher by not handing in that work.

Many students fill their free time with other interests. "I'm always gone skating on the week-ends and at the tanning bed after school. I like to do FUN things."

Time management is a problem. "I have too much to do. I run out of time." "My parents expect me to do so many extra curricular things: cheerleading, gymnastics, sports, music lessons, church groups, job . . . I'm never home long enough to get it done." "I need enough time to finish once I start. If I don't have that much time, I don't start."

"I do my homework in my other classes before it's due. This works OK except for my first period class. I have trouble getting that homework in."

Then there are those students who've learned one lesson well . . . how to slap the blame on somebody else. "I forget about it until it's too late. Someone should remind me."

"I don't feel it's very important. Nobody at home cares whether or not I do it."

And, of course, that old standard, "Teachers give too much

homework"

So, what drives those successful students, many of whom have interests outside of school? How do they manage to get their homework done?

Most have a plan. "I will write down in my planner what I have for homework and then I'll take these books home. Whatever books I bring home, I know that those are the subjects that I have homework in."

Some students have found what works for them. "I try to do my homework as soon as I get home, unless I have some extra-curricular activity" or "I do my homework about an hour before I go to bed."

"I write it down. I do it the night before it's due as long as it is not a big project. When I have a big project, I chart it on a wall calendar and work on it a little every day. Teachers usually give you enough time, if you'll just pace yourself."

Many have immediate goals. "I have to have good grades to play basketball." Some are already looking farther ahead. "I want to go to college. I know I have to get good grades and prepare myself." And many have parents who have made their child's schooling a priority. They check, they praise, or they punish. Kids don't want to get grounded, which looms as a real possibility.

One young man put it philosophically. "People who don't do their homework are really kind of dumb to me because I don't know of anyone who would want to take the same grade over and over and over and over."

He's figured it out.

Haunting Former Students

We were in a local sports shop the other day. The clerk looked familiar. He waited on us and helped us get the correct basketball shoes. Then, as he was handing me a pen so I could sign the credit receipt, he looked up sheepishly and said, "You know, Mrs. B, I still owe you a paper."

It seems he missed the last few days of my class and never turned in the final assignment. "They did let you graduate, didn't they?" I was joking.

He was not.

"Yes, they let me graduate. But, I still owe you a paper."

This little scenario convinced me of something I've believed for a long time...teachers continue to haunt their students, long after they've left our classrooms. Sometimes the memories are pleasant; sometimes we roam in their nightmare lands.

I absolved him of his crime and handed the pen back. I wouldn't have been surprised if he had insisted on bringing the now moldy paper for me to read.

Once we get out in the work world, we've all realized that we could have worked harder in school. We learn lessons about how hard you must work to make a living. It is then that school days seem carefree.

But, there seems to be a lot of guilt involved in whatever effort we chose to make.

Whenever I meet someone for the first time and he learns that I'm an English teacher, I will get one of three responses:

1. "I always hated English." (math types who loathed discussions of the various meanings of metaphors "What do you mean white can be death, white can be life, white can be anything?"

2. "Oops! I better watch what I say." As though we're on constant grammar patrol, rooting out the errors we hear. Are we really

that obnoxious? Maybe we are.

The most interesting thing comes most often.

3. "I've been reading...." And they'll give me an impromptu book report. They want to know if I've heard or, better, read their book. They'd like me to bless their extracurricular efforts. Perhaps they hope there's a web page for English teachers and I'll issue a report to their old high school literature instructor.

We can all relax and stop feeling guilty for what we did or didn't do in high school. We all come to realize that we could have done better, we could have done more. But we didn't. We made other choices.

Like many parents, I want my children to do well in school. I sometimes set expectations based on my wanting them to excel. They don't always meet my standards.

But, spend time with successful folks in the adult world, people who are good at their jobs and happy with their lives, and you'll find they weren't all straight A students.

What is reasonable is to ask your child to do his best. Spend any time with him as he does his school work and you'll get an idea of what that is.

He needs to work to his best ability. He will need that for the work world. He needs to develop pride in his best effort. That will motivate him through school and on into life

Homework Place

"Mom!" he screams from his bedroom. "I have to bring in my volcano tomorrow."

Volcano? What volcano? You mean the one we talked about? The one you're going to make with papier-mâché and rig so it will erupt? The one that is 50% of your grade? That volcano? Tomorrow?

And gritting your teeth, you stomp to the garage, slam the car door, fire up the engine and zoom out of the garage. Mutter, mutter, mutter. How did this happen? What about all our planning? Thank goodness for late-night mega-markets.

Schoolwork can reap havoc on your quiet little home. Whether it's the daily stuff or those big projects, it means hassles, mess and time. As a parent, you know that it is important and you need to be ready to help out, but how can you minimize the stress homework causes?

Study experts suggest that you and your child find a homework place and dedicate it to that activity.

School kids gravitate to some common sites in the house for homework, according to an informal and highly unscientific survey I recently took:

The Kitchen Table—often the first resting place for books, so it's convenient. You as parent can keep your eye on the work and can help, but this most busy room holds lots of distractions, like the refrigerator and supper aromas.

The Couch—after a long day at school, even your youngest wants to put his feet up. But, this is way too much comfort for the student who needs some serious study. Also, usually there's no writing surface.

The Big, Soft Easy Chair in the Television Room—for many of the same reasons: No, no, no. Bad choice. Television is just too enticing. Your child may tell you he can study with the TV on but trust me, he can do better with it off. His eyes need to train on his work.

The Bed—no one sits up on a bed. It's too easy to flop onto the belly and then it's only a few deep breaths to dream land.

A good homework place is a relatively quiet spot, away from distractions, and comes with supplies.

A word about quiet: some students require absolute silence to concentrate. Some actually need some background noise to do their best. It blocks out distractions. The noise needs to be something they're not really listening to, like instrumental music. It can be the hum of the dishwasher or the whirl of a fan.

Do whatever you can to minimize distractions. In an ideal world, you might unplug the phone. At least, have supplies nearby.

Or: Johnny is working on his mathematics. His pencil breaks. He has to go look for a new pencil. Roaming around, he stops for a snack. He finds a pencil by the phone. The phone rings. It's his Little League coach. He writes a message to his dad about practice. Thirty minutes later, he gets back to the desk. And, oh yeah, where was he in the math book?

Once you've located the homework place, stock it with supplies. When a child needs to walk around to find something, he clips his concentration and must regain it.

Besides a table, chair and a good light, here is a list for the well-stocked homework place:

> BASIC: pencils, eraser, and sharpener, notebook paper, pens, white-out, assignment book, calculator, dictionary, atlas, glue, scissors
>
> SUPER: markers, colored pencils, ruler, tape, stapler, encyclopedia
>
> DELUXE: construction paper, paper clips, index cards hole punch, rubber bands

Nearby keep a few pieces of poster board and a stack of old magazines.

Your child may have his own computer and this could be a luxurious aid for some homework. However, I'd put the PC where I could monitor it because there's a lot of fun stuff he can do that has little connection with schoolwork. "Blasting the Aliens of Zarno" is more exciting to a fifth grader than reading about Benjamin Franklin. Go figure.

One quick shopping trip can take care of supplying the homework place. You may still get to make that late-night papier-mache run some day, but tell the volcano maker it'll cost him.

Teachers have lots of ideas about homework. Homework is a hassle for students but a parent's attitude can make a difference in how he approaches his task.

~ ~ ~

You see them every day at the same time. School's out and here they come, one arm lower than the other, lugging a book bag. Home they go to begin homework.

Whatever happened to those carefree hours when the teacher lets them go?

Kids need some leisure. On the other hand, the jobs of the future require better preparation beginning at an early age. We're expecting more of students; homework is a necessary element of after-school time.

The homework debate rages. Should there be some? How much? At what ages? How often and how long should it take? Three opinions are common:

I. SCHOOL IS LIFE—kids should have homework every day, even on the weekends. They need to develop the habit when assignments are brief so they'll not go into shock when homework requires big effort. The whole family should revolve its life around the student's need to do homework.

II. NEVER, EVER, NEVERLAND—kids need to play. Kids need to play after school. Parents make plans without checking the school calendar. After all, the kids may belong to the school from nine to three but after that, it's party time.

III. HIT OR MISS HOMEWORK—If the kid didn't finish at school, he has homework. It's his punishment for working slowly. Or, it indicates that the teacher didn't plan well. Or the class goofed off. Either the student gets homework because he messed up, or the teacher becomes like Cinderella's wicked stepmother, who always found one more job for her to do. Scheduling after-school activities is a point of turmoil. And it's all the school's fault.

None of these views works and each has negative effects on learning. If there's to be a benefit to homework, we need a positive stance.

What's the best way to see homework? What is it supposed to accomplish?

Think about it: learning has been a constant activity for your child. A baby sees, thinks, touches; then he turns away, returns, looks again and then files the information away for future use. Formal schooling should follow that model.

For example, the math teacher introduces "least common denominator." The student reads, watches the teacher and maybe works a few problems. He sees, thinks, and touches pencil to paper.

Then it's off to another class, lunch, or the playground. He turns away.

Later at home, he pulls out that book and returns to the concept. He works some problems, checks them and either feels confident or knows he has to ask questions tomorrow. He files the information away, ready to retrieve it the next day.

So, homework can be reinforcement for the student. He can discover whether or not he's "got it." And if he hasn't, he knows to ask the teacher.

Homework time should be scheduled as righteously as any other after school activity.

(My students moan) Classic guidelines measure homework time as ten minutes a day per grade level, up to one-hundred minutes. So a second grader should put in twenty minutes; and a 9th grader plans for ninety minutes on average.

The average is important because there are those big projects that take hours and hours on days and days. That time should be balanced by shorter study times on some days.

There should never be a No Homework Day. After all, school is your child's job for the next few years. But that doesn't mean teachers need to assign something just to fill time.

A child can always reread a portion of the social studies lesson. He can read an article in a magazine or an encyclopedia. Even if it's not directly related to what he's studying, he is using his brain and will store this information. He puts in the time.

So, help him practice the habit of homework.

Kids Need to Be Somebody

It wasn't the Academy Awards. It wasn't the Golden Globes. It wasn't to honor the best (sic) on MTV.

It was—drum roll, please—the county-wide competition for the best speller, grades one to eight. Parents, teachers and fans met in a big auditorium where about forty youngsters took their places on the stage. They ranged from nine to fourteen. Each represented his school, having bested all his chums to emerge as the campus champ. The contestants would spell until they were stumped. The last one on the platform was headed for a state contest and finally, national competition in Washington, D.C.

One of my favorite nieces was also one of the youngest competitors. When she had told me she was her school's winner, she mentioned another boy was her alternative. I had asked, "And how do you spell that?" Without missing a beat, she giggled and spelled it.

During the first round of competition, someone got "Alternative" as a word. She shot me a glance from her seat and we shared a smile. She did just fine in the spelling bee and although she wasn't the last one standing, her cheering section was proud.

For a group of local students, this was a chance to shine. Even if they never compete again, they have the memory of being part of an exciting evening.

In addition to teaching our students their three R's and all those other letters, schools need to work to give our students opportunities to be a part of something. Kids need to find a place within the school where they can be special. Academic competitions and sports afford that place to many. For others, schools should foster and encourage clubs and service organizations.

Just the other day, a gentleman called me by my first name and joined me for breakfast. It took a moment, but I realized my companion was a student from twenty years ago. During the course of our meal, he described the paths he had traveled to land at his current career. He's been busy.

I remember him as being a quiet, shy young man who was good at taking directions, but needed a push. When he was going through school, we weren't as concerned that everything prepared the student for a job as we seem to be today. So I had to ask, "What did you do in high school that led you to your career?"

He smiled and said, without a doubt, it was the club that placed him in a leadership position. This was, by the way, a no-glory but lots of work kind of club. We had set impossible financial goals for ourselves and then, shock of shocks, we had met them. Our little accomplishments went unnoticed by most, but they remain a point of pride for this young man. It gave him confidence to pursue his career choice.

If you ask people who have been out of school for a while to recount a favorite memory, it may be one class or one teacher, or a club activity.

How many of our students are hungry for just such a place where they can feel special? Many. We need to remember that.

Kokomo High School Latin Club, 1920, meets outside the main entrance for a group photo.

A Common Disability

As a society, we've identified many areas of legitimate disability and have decided to offer help. This comes in the form of social security payments, close in parking spaces, special education opportunities, and patience that grows from our collective awareness.

In schools, we've saved many children that struggle with physical, emotional, or cognitive disorders from failure .

But there's one disability that we've not found a way to remediate. We see it demonstrated in school, but have few ways to address it. It will cost our society dearly as it manifests itself in the workplace. Its acronym is L.A.Z.Y.

We teachers have the audacity to think that our students should do some work away from the classroom. We think they should prepare themselves for the lessons of the day. We think they should pump a little mental iron as they move through their courses of study.

But, some of the bodies who fill the seats have different ideas. They have decided that school is just a brief sentence. They'll serve their time and then get out. Don't expect them to give it a thought when they get sprung through the door at the end of the day.

A teacher recently assigned her students to gather leaves of the various trees in our state. She wanted them to learn about these plants in a quality, hands-on way.

The students were to display their leaves on poster board and label them. Not a hard task. It would just take time to gather fifteen different leaves and get them catalogued.

How much time? You could probably figure that out. These students had three weeks to complete the task. They used class time to hike outside to areas with many trees. They didn't even have to risk life and limb climbing up to branches either, since it was autumn and the leaves were easy to reach. Of course, they had to bend down a few times.

While many of the students faced the project as a fun challenge, there were those who just wouldn't do it. These kids came to class empty-handed on the due date. They had worked on their excuses. They whined that they hadn't had time. They whined that they couldn't afford the poster board or that they couldn't find any glue. They whined that it's unfair to expect them to do work outside of class. All that hot air! They might have directed their energies at making a token effort toward completing the assignment.

This is merely one example of what goes on every day in most classrooms. And, believe me, teachers have scratched their collective heads, trying to find ways to motivate these LAZY students.

You might suggest that the teachers should weight an assignment so that a student's grade would be in jeopardy if he doesn't do it. You probably did your leaf assignments in school. LAZY students don't care if the project counts 10%, 50%, or 100% of the term. It doesn't matter. They're not going to do it.

But they will be out there some day, looking for a paying job. And when they get fired for their LAZY problem, they'll whine all the way to the unemployment office.

Chapter Five

Stories From School

Like most veteran teachers, I have lots of stories to tell. I've seen lots of students grow from funny freshmen to serious senior, ready to take on the world. And, while I've presented a curriculum, I've learned from my students lessons about what it takes to succeed, in good conditions and in spite of bad situations.

There's no one central truth about school success, but caring adults make a huge difference in the lives of kids. When kids know that someone cares, they will try to reach for their best. We owe that to our children and to our students.

One Becomes an Adult

When it comes to parenting, there are three really good times to offer advice: when you are a young, childless adult; when you've gotten so far past parenting that details are dim; and of course, during your teen years when you keep making lists of all of your parents' mistakes and all the things you're never going to do when you have kids.

People down in the trenches with actual children don't offer too much in the way of "this is how you should do it." They know, especially if they have more than one kid, that theory may read well but what works today may flop tomorrow and last week's solution is old news. Parenting is an inexact science at best. What we can hope for is the raising and launching of fine human beings into the world.

Since I'm in the middle of raising kids, I'm not ready to write a book on the subject, although I could conduct an empirical study. I face parenting successes every day in the students in school. What are the secrets that help one child sail through his adolescence while another struggles all the way? Why do some kids charge toward challenges with confidence and steel themselves against failure while others procrastinate and fear defeat? I'm sure that temperament has something to do with it but I'm betting that some really alert parents work hard as their kids grow up.

I'm no expert. However, one of mine is turning twenty-one, and as I parent less and be a friend more, I can reflect on our journey to her adulthood.

For her parents, it's been a smooth trip. First, we were blessed with an easy-going, healthy child. She joined our family when we had grown up a bit and had some money in the bank. We led her and she gravitated to friends and adults who affirmed her talents or taught her new things. Also, she found her way, guided by relatives, pastors, teachers, coaches, and neighbors. She grew from a child who followed our lead to a young woman who chose her own path.

In her case, I think the best gift we gave was that of stability in her day-to-day life. That meant routines and schedules.

There are lots of kids who handle upheaval well, but that tedious same old same old every day helps many kids concentrate on schoolwork, family relationships and personal growth.

That stability includes a long-term marriage, but let's face it. We aren't the Cleavers, Seavers, or Huckstables. She has witnessed the down and dirty of real life. We've sat with her through sermons on Christian values and then brought her home honed to witness our falling short.

Some would say we've all been lucky. I disagree because I don't believe in luck. I believe in a power bigger than myself. God is the perfect constant, the same "yesterday, today, and forever."

My daughter has several dedicated people who pray for her every day. Whether they know specific needs or not, they set aside daily time, asking for guidance and wisdom for her. Without preaching, they have let her know this quiet constant.

My little girl is now a legal adult. Thanks to all who have helped us raise her. And to those faithful on-your-knees folks, keep praying.

Organization System

We recently had one of those "cookies and punch" things in the main office at school. I'm not sure, but I may have missed Teacher Appreciation Day. It doesn't show up on all the calendars.

Although teachers like the goodies, that's not our major perk. We love to know we've made a difference in some student's life. So, I need to thank a teacher, and I'm not sure to whom I should direct this. Someone lit a fire under one of my kids.

There comes a time when our children start listening carefully to someone other than their parents. Often, it's to their teachers. That's one reason it's important who your child's teachers are.

I have had, at home, an organizationally challenged child. We've found it especially difficult to transport papers, letters, and homework to and from the school. Too often to count, I have found important things crammed into the jacket, crumpled in the bottom of the book bag, or stuffed into pockets before or after washing.

All this changed at the start of second semester. He came home and announced that all of his grades were going to go up and he was going to make the honor roll. "Oh," I said. "How are you going to do that?"

He said that he had a new system. He needed pocket folders for each of his classes. "I'm going to put all the take-home papers on one side," he said, "and all the homework papers on the other side."

He then whipped out his planner which, at half way through the school year, looked quite new. "I'm going to write down all my assignments in each class every day." He showed me how there was a space for every subject. "Then, when I get my work done, I'll check it off and put it in the right folder."

He continued, genuinely impressed. "I'll put the folders back in my backpack and hang it by the back door."

What a great system.

I drove him to our local supply store and he took his time choosing folders. He fingered the stock and pulled one down. Then he examined it, opening it and stuffing his hand into the pockets to make sure they were sturdy. He handed it to me. He repeated the process, again and again.

He knew what he was looking for. He wanted a different color for each class and didn't want pictures of kittens or unicorns.

We finished our shopping with some fun pencils, my only contribution. I didn't even ask him what had happened to all those other folders I had bought for him at the beginning of the year.

He put the system into practice and so far, so good.

What's amusing is that there's nothing profound about this organizational model. I've been preaching it for years, at home as well as at school. But, apparently my voice has become a hum and he didn't hear me.

A new preacher stepped up to the pulpit and this time, my son was listening. He became a convert to pocket folders and planners and is turning in his work, on time, to the right teachers.

I'm grateful that he was listening, and I'm grateful that a teacher decided to go over it again.

Reading to Kids

Last week in class, we finished our study of the short story. To prove to them that they had indeed learned, I pulled out a childhood favorite so we could apply our lessons.

The book's bright orange cover gave it away before I showed them the title. They knew this book. They'd sat on someone's lap long ago and listened to the story. Then we read Dr. Suess's *Green Eggs and Ham* and discussed setting, plot and conflict. Smiles all 'round.

Lots of good mommies and daddies read to their babies at bedtime. That's a special time and it goes by so quickly.

At our house, we started at six months, as soon as the critter could sit up. A friend had passed on six *Golden Books*. I wrote my daughter's name in each one and set them up next to the maple rocker by the crib. She especially like *The Pokey Little Puppy*. We read it until it dissolved at the binding. By then we had the whole thing memorized.

Amateur dramatics, cuddle time and excuses to ignore mundane chores. That's what reading to children is.

Pleasure is a plus. On top, you're teaching the baby the sounds of language, the structure of sentences, and proper grammar. And you're proving that stories are a joy.

Then, somewhere most of us lost it. Maybe it was a new baby or a new job, maybe some family upset or moving to a new home. Before you know it, the child can read and after a few reads to you, he gets the idea that you'd rather do something else than listen to a story that may keep a six year old on the edge of his seat, but it's not to your liking.

Besides, there's TV and other interests. Those chores, you know.

In simpler times, the family gathered around the hearth and Papa would read. It was the evening entertainment. In some ways, these were richer times filled with poetry and long tales that held the children captive as Mom darned stockings.

If you can get the crew to slow down, you might try rediscovering the pleasures of reading aloud and having someone read to you.

When a friend bought her family van, she opted out of purchasing a VCR that plugs into the lighter. Instead, she uses long family trips as an opportunity to read to her children. They are, for a few hours, a captive audience. She's taken them through the classics with a leisurely pace of someone who has eight-hundred miles to go. Even her older children, who are teens, unplug themselves from personal sound systems and find their way to the back of the van to listen to Mom.

If you get a chance, you might try to reestablish this habit at home. Start with some sure winners, like C.S. Lewis' *The Lion, the Witch and the Wardrobe*. It's a fantasy tale that captivates children. Lucy and her friends hide in a magical wardrobe filled with fur coats. As she feels for the back of the closet, the fur coats become fur trees and she is in a forest. So the adventure begins.

Older children and adults can easily see that this story has deeper and more complicated meanings, which makes it so satisfying for all readers.

Your local librarian can make other good suggestions for oral reading.

You don't have to be a great actor to read aloud. You may be a little rusty, but with practice, you'll do fine.

You may even start a new family tradition.

Rising SATs

National scores on the Scholastic Aptitude Test have been a barometer of academic performance for several generations. Every year, we read reports comparing the scores of seniors in this school or that school, and always, we see a comparison with scores from thirty years ago.

The reports are often dismal and critics step forward to point index fingers at the public schools.

We could analyze the academic status of students who took the test in the past. That might give us a better comparison. We didn't insist that everybody needed to go to college, so only students in the college-prep tract took the test. In the last twenty-five years, too many have pushed college on students who have no business there. Since the SAT is an admission requirement, many more students take the test, including those who have no preparation for the academic rigors of college Their scores get averaged in and the overall scores go down.

The news has been a bit cheerier lately . . . students are choosing more difficult courses in school and SAT scores are beginning to creep upward. Armchair educational critics are wondering: could there be any connection?

As they say in the hallways of your local school, "Well, DUH!"

Early in my career, Indiana required six semesters of English to graduate. The only students who took English in their senior year either loved the study and wanted the challenge, or they had flunked required courses and needed to pick up the credit.

I taught one of those fill-up-the-credit courses, and some of the students were college bound.

This course was designed for students who were weak readers. We read several fine books that had vocabulary and difficulty indexes below 9th grade reading. We need courses like this for students whose gifts lie in areas other than English, but we permit students to choose and my students, many of whom could handle a challenge, chose the easy way out.

They enjoyed the joke. They could breeze through the books and sit and grin while others struggled.

The joke was on them, however. They certainly didn't pick up any skills that would help them in college. They took the SAT and their scores figured into our school's average.

Today, although there are easier paths, many college bound students are choosing to challenge themselves with Advanced Placement and other difficult courses of study. Their curriculum requires them to read mind-stretching, world-class works of literature, history, and philosophy. Then, they discuss and debate the ideas and write lengthy papers, exploring their understanding.

Is it any surprise that students who feed their brains with high-test fuel are able to produce quality results on standardized test? The SAT requires students to recognize and use advanced vocabulary. They must also demonstrate their grasp of complicated ideas and their ability to consolidate their learning from various disciplines. Those who have explored a wide range of challenges have more from which to draw.

If you want your child to succeed in the academics of school, to rack up an impressive SAT score, and to forge his way through college, you must impress upon him that hard work now pays off. This is true in school. It is true in life.

Where Do Rude Kids Come From?

When my children were learning to talk, one of their first phrases was "cuse me." They picked this up from their mother. It's not that I'm overly polite, but I am inherently clumsy. I'm always tripping over feet and bumping into people. That's my manner from way back, so saying "excuse me" is as much a habit as breathing.

I was thinking about this the other day when I was reading about a survey in the newspaper. Public Agenda, a research and education group, conducted phone interviews with 3200 adults, many who identified themselves as having a lot of contact with teenagers. The results are a report called KIDS THESE DAYS, commissioned for Ronald McDonald House Charities and the Advertising Council. The participants spanned social, financial and racial lines.

Sixty-one percent of the adults in the sample think it's a serious problem that children don't learn such values as honesty and respect. Eighty-one percent of parents think parents are not being good role models or that they fail to teach their children right from wrong. sixty-five percent describe teens disapprovingly. Respondents use terms like "rude, irresponsible and wild" and say adolescents "lack discipline and are spoiled."

I had time to read this as I stood behind a well-dressed woman in the express check-out line as she loaded her cans, bottles and boxes onto the conveyer belt, underneath the printed sign, "This line for twelve items or less." She had no trouble counting out her cash, so I assume she was aware that she ran twenty-four objects by the checker. Did she break the law? Of course not. She merely thumbed her nose at the other shoppers.

What's the motivation? Does ignoring common courtesy give her some satisfaction? I don't know.

After I bought my five items, I walked out to my car. In the parking lot, a shiny pick-up pulled into a handicapped parking space and two healthy-looking men got out. They jumped down, slammed their doors and bolted for the store. They appeared quite able. Maybe boorishness is a disability.

I turned onto the highway and started home, driving at the speed limit. About a mile down the road, a man with one of those "Kids First" bumper stickers rushed up behind me, illuminating my rear-seat passengers. He flashed his headlights and then sped around me, crossing the yellow line and cutting me off. Then he tore off down the road. Maybe the kids in his van needed him to get somewhere faster than the law allows.

Ok. I'm done venting. It was one of those horrid, humid summer days when tempers run short. But my point is that there's a lot a rudeness in our society. While some children are left to fend for themselves, many are tutored carefully in the fine art of Me First by their parents and then by the rest of us.

Most of the people I know who have significant contact with teens are teachers. Most of them like the students they work with. They'll tell you, almost all of them, that kids live up to the standards you set. Or they sink down to them. Teen-agers can be rude, sullen, even scary. They can be also be witty and cheerful. My favorite adolescents make me laugh at the absurdities they see in the world.

Some of the folks who took part in the survey see the possibility of solutions in public schools, religious groups, and community involvement. And we could all try to model those positive behaviors we'd like to find in tomorrow's adults.

Simian Lunch

You haven't lived until you've broken bread with a booth-full of middle school boys. This was my privilege recently, when I took a batch to get a basketball autograph in Indianapolis.

I'd usually not be welcomed to their lunchtime banter, with my status as mom and teacher, but then, I drove.

Some anthropologists have suggested that early adolescence is a slide back on the developmental scale. Where once stood a recognizable human child now slouches a giddy mass of constantly moving flesh. What comes from its mouth is often unintelligible, if not unprintable.

They are a fascinating combination of emerging adult and giggly child. Even in big bodies they are, for the most part, still little boys. Unencumbered by social inhibitions, they will tell you everything.

We opted for a metal tableware restaurant. They displayed shrewd menu savvy and each requested a separate check. "And extra straws, if you don't mind," they said to the waiter.

They needed the straws for adornment: up the nose, in the ears, and as spit wad pipes when I went to wash my hands.

One grabbed the others' attention with, "SHHHH, SHHHH, SHHHH..." then a little, um, wind. Hoots and screams back and forth to each other. "Oh, gross." "Gross, man." "Whew." Then a detailed discussion of texture, comparison to other odors, and the retelling of the great gas passes of the past. Who says we don't teach higher order thinking?

You may remember: a middle schooler will scratch his chin by snaking a hand up from the belt, under the shirt, baring his belly for general inspection. He talks to friends as if they are outside, not two feet away. The hostess, displaying great wisdom, had seated us off by ourselves.

Most conversations I have with kids roll around to school. I say, "How's school going?" They say, "Fine." It's what we all expect, we smile and then we go on.

Not my lunch bunch. They seized upon our temporary camaraderie as a chance to vent. They had rehearsed their stories with each other and now they have a new audience.

These young men (?) would like to spend more time playing basketball than sitting in a classroom. The lessons drone on and on too long for their sporadic attention. So, to keep some focus, they construct stories.

"Mr. Jones owns one pair of pants. No really, he spilled ketchup on the leg on Monday and wore the same pants all week. And they don't fit. His belly hangs way over his belt."

"We like to walk down the hall behind Miss Smith. She is kinda fat and wears tight black pants. She puts money in her pockets and the coins work their way up and fall out on the floor."

They have documented every time a teacher had dandruff, wore snagged hose, scratched his nose, lost his glasses, dropped the chalk, or blasted them with coffee breath.

They're extremely observant. And they remember. One wishes they could train the same powers on their studies.

This is a critical age. They gripe about these adults but also about each other.

My little group had much to say about their teachers, much of it negative. They affirmed each other's tales of woe. But, they also hold some teachers in high regard.

Mostly, it's those who can muster compassion for these wiggly worms.

King Solomon, the wisest of the wise, said, "A soft answer turneth away wrath." It also can charm a middle schooler. It can get him to do his English and to study his math. Society is cruel to him. Nature is cruel to him. He is starving for some kindness.

It takes special gifts to teach in middle school. And, sometimes, it takes the wisdom of Solomon. If you send one of your children in there, keep that in mind.

A Teacher's Many Hats

Tell me again what a cushy job I have. Roll your eyes and say how you wish you got paid in the summer for not working. Or that teachers can't be fired and can sit back, pass out a few worksheets and call it a day. Or maybe you'll take the other tack. You'll tell me how you wouldn't do my job for any amount of money because it's so hard and I must be so dedicated. Then, when I'm gone, chat with your friends about my soft, secure position.

I don't want your sympathy. Like most teachers, I love what I do much of the time and like it most of the time. I wouldn't stay in it if I didn't. But, there's a lot more to what I do than teach a subject.

Whenever a critic slams the schools for what we're doing wrong, I long for a job description that I can print up, stare at, and figure how to fill the requirements. I know the hours my contract covers. I could sing you that old song about how teachers put in lots of extra time. It's true, but I won't bore you. I have a thick handbook that tells me how to function within a certain building. Pink passes, blue passes, nurse lists, field trip forms, accident reports, discipline slips...procedures differ from school to school, but it's all there in black and white, if I'll just check the table of contents.

There's a lot I can do on auto pilot. But, nothing prepares a teacher for cultivating the daily relationships with those special students in his classroom. Each child brings unique gifts and also weaknesses to the room. He also carries his burdens and his joys. Sometimes life is good and he wants to join in the learning. Other days, woes weigh him down to where he can hardly lift his head.

Beyond the regular duties, teachers get summoned for service beyond the call.

Just the other day, a girl stuck her head around the corner as I was packing up my briefcase. She needed to talk; I needed to leave...I needed to stay.

She settled into a nearby desk and complained that her new braces hurt. She wanted me to see them and she wanted me to tell her that they'd hurt less in a few days. She already knew this, but she needed someone to say it.

She then asked what she could do to pull up her grade in another class. She was knocking heads with that teacher. She said that every time she raised her hand to ask a question, the teacher would yell at her and then answer the question.

I told her that it was good to ask those questions and that I understood how upsetting the yelling was. So I suggested that before she raise her hand, she remind herself that the teacher would be yelling. Then, at least, it wouldn't catch her off guard.

Finally we got to it. Looking at the floor, she told me that she was now living with her father because it "didn't work out" when she lived with Mom. She raised her head and looked me straight in the eye. A tear welled up and ran down her cheek. "My mom doesn't want me," she said.

The uninformed tell me kids are tougher than that. Right then, I needed to be Mom and I needed to want her. She needed that more than language arts.

Are you uncomfortable with my stepping into your shoes? Then don't let me.

What Mother Wants

Two weeks before the Big Day, a series of commercials appeared on television. Moms of assorted sizes and ages were asked what they wanted for Mother's Day:

"Oh, I don t know."

"Whatever you pick out is just fine, dear."

"All I really want is a hug."

These short messages sparked a shopping frenzy at the mall on Saturday afternoon. Sunday morning at the grocery/lawn care emporium mega-mart, any goo-gaa with a bow was a sure sale.

Later on Sunday, Mom gathered the family around and OOHed and AHHed over her bounty. The givers felt foolish. How do you honor this woman whom you've given the label MOM?

In the most guilt-inducing commercial, one forlorn lady sighed at the camera and said, "Being your mother is gift enough."

YUK. YUK.

But, wait a minute. Isn't that the truth, Mom? Have we sunk so low that even motherhood has a pay-back tag?

Ask the young woman who cannot have children. Ask the wife as she returns from the grave of her newborn. Ask the late-married lady who finds she missed the age requirement for adoption. Ask that favorite maiden aunt, the one who showered you with gifts every time she came for a visit. Many a woman would give up an appendage if she could trade it for a child of her own.

When you find yourself moaning about the dentist bills, the price of shoes, thickness of a teen's bedroom floor, the noise, the mouth—somewhere close by is one who would trade her less-troubled life for your misery.

Most moms don't have to fake delight at the burnt breakfast, the homemade hair bow, the hastily purchased cologne on Mother's Day. They understand that their job description includes frustrations along with the perks.

Teaching school is like that. We pick up the ball from parents and try to keep it rolling. One hour a day (or 1 1/2 hours every other day) we create a home for your children, with goals, expectations and learning.

You give them meals, clean clothes, moral training and love. We give them knowledge, discipline and life skills. We continue socialization on the foundation you've built. (No, Johnny, it s not ok to belch out the alphabet.) And love? Yes, many of us love many of them, sometimes when no one else will.

Why do we do it? For the perks? Well, we've all receive our share of sample cologne, funky ties and homemade marshmallow concoctions. But, that's not it.

As teachers, we work for the AHA!

We orchestrate the lesson. We draw, we diagram, we illustrate, we lecture, we question. And, if it all clicks, we get to see the student's eyes grow wide, the smile beginning to break through, and the glimmer of that light bulb going on inside his brain.

Or, we reach out to the terminal pouter—that dour one who hates life, school, and especially teacher. We work to bring him in, with the fervor of tent evangelists. Our message is that learning is the key to a happy life, at least the here and now. And we wait for him to answer the altar call. Sometimes he does while we have him. Sometimes it takes years. But, most of us hold to the faith that he will see the light.

Of course, each teacher has a special drawer where he keeps those notes of thanks from parents. Those mean a lot. And, a perk of a long career comes when a former student, someone who has gone far away, takes a moment to remember that one who tried to light the spark in his mind.

Those pieces of paper are the most valued gift a teacher ever gets.

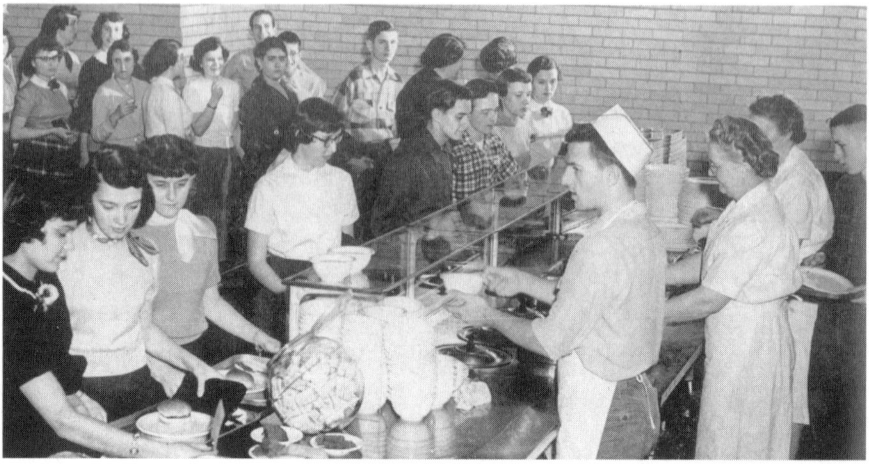

Some things never change. Students love lunch. *1953 Sargasso*

1955 Sargasso

Planning for the "important career as homemaker" meant learning how to sew.

Chapter Six

Final Word

Thanks, Parents

Teachers have mailboxes at school. These get stuffed with an interesting assortment of messages, announcements, and forms. Many come in bright colors, so on any given day, the opening looks like a fluorescent rainbow.

Then there are the do-dads. Recently, I found a very nice little desk clock in my mailbox. It requires a battery...there was no battery. I don't want to sound ungrateful; I'm not. But, who has time to shop for an odd sized power source at this time of the year? It may be months before I know if the clock actually works, but it's a pretty paperweight at present and that's just fine.

Last week in my mailbox, I retrieved what I prize the most: a personal thank-you note from the parents of one of my students. They took time from a life that is at least as busy as mine to express their gratitude for my work with their child. This particular student is one of those delightful kids who loves school, loves teachers, and so is easy to teach. It took little effort on my part to motivate her, but I appreciate the note, just the same.

Teachers get letters like this only once in a while. Most of us realize that for every note we get, there are twenty others that someone meant to write. We don't work with kids for the strokes, but they are special, just the same. I know some rather unsentimental teachers who have carefully preserved every parental thank-you they have received over the space of their careers.

This is a good time for teachers to return the favor, so here's a thank-you note to the parents whose children walk the halls of the local public school.

Thanks for nurturing that infant, that toddler, that pre-schooler. When you trained him to behave, to respect authority, and when you encouraged him to explore, you gave him important skills for school. Every time you read to him, starting before he seemed to understand, you not only developed a special relationship with your child, you introduced him to his most powerful educational tools: words and ideas.

Thanks for making sure your child had a healthy home life: good food and a regular bedtime. Thanks for requiring him to take on some household chores and for insisting that he complete them in a timely manner.

Thanks for getting him up and to school on time on a regular basis.

Thanks for setting aside a place for him to do his homework, and for taking the time to check to see that he was doing it. Thanks for being available to him when he needed your help.

Thanks for letting him stay home when he was sick, even though it complicated your life. Thanks for making him go to school when he faked illness, because you knew he had forgotten his homework.

By nudging him to be responsible, you've helped him do his best in school. He'll not disappoint you.

Thanks for loving your child more than anyone else can.

Endorsements

She writes like I think, with common sense and honesty. Everything she writes makes sense; she's very sensible.

May Jane Carey of Kokomo
mother of 9 and grandmother 12

Lynne Bolinger offers an insightful, conservative, and humorous view for parents and teachers. Kathleen Parker and Chris Matthews . . . stand back.

Dick Evans, parent and potter

Hurray to Lynne Bolinger and her column, "Learning no means no." It annoys me that criminals blame the system for their problems. We all have choices and they made theirs. I've seen first-hand what she's talking about and I agree 100%.

Kokomo Tribune, Sound Off 11/13/01

I compliment Lynne Bolinger for her column about saying no and raising your children . . . Right on, Lynne. The rest of you parents, get with the program

Kokomo Tribune, Sound Off 11/14/01